WHERE ARE ALL THE ALIENS?

Terri and Melanie Pierce-Butler

Where Are All The Aliens

Copyright © Terri Pierce-Butler, 2007

All rights reserved
ISBN 978-1-4303-2260-3

Contents

Introduction   5

Chapter 1   What Is Life?   7

Chapter 2   What Is Life Made Of?   19

Chapter 3   The Solar System.   41

Chapter 4   The Earth.   61

Chapter 5   Life And The Rocky Planets.   87

Chapter 6   The Gas Giants.   107

Chapter 7   The Sun And The Solar System.   123

Chapter 8   The Sun As A Star.   137

Chapter 9   Planets Beyond The Solar System.   155

Chapter 10   What About Intelligent Life?   173

Chapter 11   Epilogue.   191

Where Are All The Aliens

Introduction

One of the most interesting questions so far left unanswered is whether life exists, other than on the Earth. The discovery of just one, provable, extraterrestrial microbe would revolutionise much of biology. The discovery of an extraterrestrial intelligence would revolutionise not just science but the whole of our culture. The possibilities are endless and exciting and include the true globalisation of our society. Naturally our culture, in the way we understand it, may very well be damaged by contact with an alien technological race. If we look back at Earth's history there are many instances of whole cultures and societies being subsumed by the introduction of a 'superior' (or at least different) technology, so we must guard against this and try not to be too enthusiastic when greeting new races!

This book is about the search for *intelligent* life outside of the Earth. This may seem a bit presumptuous bearing in mind that we have not detected any life at all outside our planetary boundaries, but I would argue that we have enough data to take a stab at the problem. Very recent advances in our understanding of the cosmos coupled with an increase in knowledge of how the Earth functions means we can narrow down our search enormously.

What I mean by intelligent life is really an alien race that is able to communicate, or possibly travel, between stars. This may seem to be an arbitrary definition but, when you think about it, life that cannot communicate or travel between stars can have no effect whatsoever on our society. We couldn't even attempt to find them, at least in the foreseeable future.

What I will not do is resort to statistics, which is often no better than casting runes. I have a very broad science background and so do not have any particular hobbyhorse; hopefully this will make for an easier read. You will not find a single equation in this book, which, I am sure, comes as a welcome relief to most, though I have resorted to boxes to explain key concepts. Obviously, if you know about the concept, you can miss out the box. Where I do mention numbers they are strictly necessary, well attested and explained in the text. Feel totally free to criticise anything written in these pages, that is, after all, one of the great pleasures of reading, also of course the basis of scientific method.

Chapter 1

What Is Life?

Nowadays most people realise that not only is the Earth not flat, it is also probably not even the centre of the Universe. This observation leads to the obvious question "Is there anyone else out there?" This question means, of course, "Is there any other intelligent life out there?" So, if we came across intelligent life would we know it for what it was?

In order to investigate intelligent life, it is necessary to consider life in general. Intelligent life has to be highly organised and complex and therefore cannot arise spontaneously. To begin with it would be rather useful to consider what life is. Firstly, this question is not about the difference between whether something is dead or alive. Something that is dead clearly was once alive, otherwise we wouldn't use the word dead. In very nearly all cases we can be sure that something that is dead was once alive and cases where we cannot be sure of this generally involve microscopic organisms that have been fossilised and as such have been dead for a long time and so subject to alteration.

Of course the study of fossils (palaeontology) is very important, as it can cast light on past life that has no known modern equivalents and the way in which life as a whole has changed over millions of years. What is clear from fossil evidence is that life has existed on the Earth for at least three thousand eight hundred million years. This is very nearly as long as the Earth itself. This early life was very simple, based on single celled organisms and as far as we can tell was found in all sorts of environments. The situation remained virtually unchanged for most of Earth's history until, quite suddenly, six hundred and seventy million years ago complex, multi-cellular life arrived on the scene. From the point of view of our search for extraterrestrial life it is important to know the difference between life and those things that are not alive and have never been alive.

"What is life?" can seem to be a trivial question. I look out of my window and see birds messing around and trees that I know from memory have grown well this year. My wife's cats are clearly alive (though if they don't leave the aforementioned birds alone I might be tempted to change this). I also have a cold, which I am told is caused by a virus; I certainly believe this virus is alive, I caught it from somebody and it has got worse, which tells me it is multiplying inside me - and hopefully my body defences will react and it will die.

Not alive things around me are equally obvious. I am sitting on a chair, which only moves if I make it, otherwise it just sits there. Fairly regularly aircraft fly low overhead but are made to do so because pilots guide them. I sit in front of my computer, which relies on someone to programme it and someone else to operate it before it does anything - it is obviously not alive. This is all very easy isn't it? Well actually, no it isn't. Nobody disputes that birds, trees and cats are alive but the scientific community is split right down the middle where viruses are concerned. With regard to non-living things there is no dispute about chairs being non-living, unless they are made of living matter, as in the living willow sculptures a friend of ours makes, but this is not normally the case.

Of course a static object is less likely to be considered alive as we generally associate life with movement and growth. Perhaps the most familiar non-living things that move are ground vehicles such as cars, buses, lorries etc. We are used to getting up close to such vehicles and so, from direct experience, we know that they are made of metal and have wheels (usually we consider something with wheels not to be alive and this is generally the case except for some bacteria with rotating tails). Vehicles have never had anyone argue about whether they are living, not seriously anyway. However, there are various research groups who are seeking to produce a computer with artificial intelligence and it has been argued that such a computer could become a living entity.

Obviously we need to have a definition of a living organism against which we can compare entities to see whether they are in fact living or not. Needless to say this isn't particularly easy and there is no consensus as to what this definition should or should not include. Three attempts at a definition are given below. They were trawled from books and the Internet and these three were the most common definitions given.

1) Life can reproduce, obtain and utilise energy, grow develop, die and respond to the environment.

2) Life is cellular, reproduces itself or comes from reproduction, converts energy and materials for its own needs and undergoes evolution.

3) Life is a self-sustaining chemical system capable of undergoing Darwinian evolution.

There is some merit in each of these definitions but also some difficulties and I believe that none of them is adequate as it stands. The first one neglects life that cannot reproduce, so the mule (a cross between a horse and a donkey, which is always sterile) would not be counted as alive. The second one's "reproduces itself or comes from reproduction" neatly solves this, though "converts energy and materials for its own needs" is somewhat vague. So many non-living things such as crystals, volcanoes, or stalactites "grow, develop and die" in a sense, which makes this statement of little value in differentiating between living and non-living things. The same can be said for "respond to the environment". The inclusion of "life is cellular" is clearly inserted simply to exclude viruses as they have a protein coat rather than a cellular wall. Fair enough if that is your intention, but then you risk also excluding the very real possibility that the first organisms may not have been cellular. The phrase "undergoes evolution" is fine to a certain extent but the same phrase could apply to almost all manmade objects as well as natural objects such as the Sun for example.

The third definition is the simplest and I believe the best of those I found. "Life is a self sustaining chemical system" is fairly straightforward but needs expanding, which I do in the next chapter. "Capable of undergoing Darwinian evolution" is much tighter than "undergoes evolution", in that it can only apply to living organisms.

Darwinian Evolution

Darwinian evolution is most commonly thought of as "the survival of the fittest". This is an over simplification and easily leads to a tautology. After all, organisms that are "the fittest" are the ones that survive to leave the most offspring, which themselves survive to reproduce; but fitness is judged by the amount of reproducing and surviving offspring that an organism produces. This argument has been used by authors of varying competence to declare Darwinian evolution as unscientific. One of these was Karl Popper, a well-respected philosopher of science, who defined a scientific theory as one in which you can make predictions that are falsifiable. A good example of this is Einstein's theory of relativity, which predicted that light would be deflected by a gravitational field. The counter argument to this was the previously accepted Newtonian physics that light, having no mass, would not be affected by a gravitational field. Einstein's theory has been tested and it has been demonstrated that light from a star was bent by the Sun. Einstein's relativity theory therefore replaced Newton's theory as the best explanation of the workings of the Universe (although because relativity is only relevant for bodies of very large mass we still tend to use Newton's theory for day to day calculations). Popper later retracted his opposition to Darwinian evolution when it was shown that it could be predictive.

The other main opponents to Darwinian evolution (especially in the USA) are fundamentalist Christians who, at present, are using the argument that life is so complex that it could not have evolved without an outside agency. This is called "Intelligent Design" in their literature, which obviously requires a designer. This is simply obfuscation and they will openly admit (except when they are in court trying to stop evolution being taught in schools) that by an intelligent designer *they mean God. There are clearly very big holes in this, so called, theory. If an intelligent designer exists then it must be a complex entity, so who designed the designer? This question just leads to an infinite regression of designers.*

What Darwin actually said can be neatly summarised:

1) More individuals are produced than can survive

2) There is a struggle for existence because of the disparity between the number of individuals produced in reproduction and the number that can survive with the available resources

3) Individuals show variation. No two individuals are exactly the same. Those with advantageous features have a greater chance of survival in this struggle (natural selection)

4 As selected varieties will tend to produce offspring similar to themselves (the principle of inheritance), these varieties will become more abundant in subsequent generations

It is clear that these four points can lead to the simplification "survival of the fittest". It needs to be remembered that implicit

> *in these points is that the environment that organisms occupy is constantly changing. Even if the physical environment remains unchanged (which seldom happens over any reasonable period of time) the environment that all organisms occupy contains other organisms, which are changing, so the goal posts are always moving.*
>
> *It has been demonstrated that Darwinian evolution can be predictive. Firstly the fossil record, although intrinsically incomplete, has in many cases shown situations where particular adaptive traits have developed in concert with a changing environment. (An example being the development of flowering plants in concert with the appearance of pollinating insects.) This means that if the direction of environmental change is known and we have a reasonable fossil record then we can predict how species will develop. Quite apart from the fossil record, adaptations for different environments have been predicted; for example island species will show dwarfism in what are generally large creatures (due to lack of resources) and gigantism in what are generally small creatures (due to lack of large predators).*

'Taking part in Darwinian evolution' does have some problems as a complete definition of life. Darwinian evolution says nothing about the origins of life (Darwin specifically made this point himself) so early life would not necessarily be defined as life by this criterion. More problematic is the fact that modern day bacteria (and viruses if we consider them as alive) are far more egalitarian than Darwinian evolution would allow, in that genes are swapped between different species as well as being inherited during reproduction. This has been a major source of antibiotic resistance in disease causing bacteria and is a cause of great concern not just for this. If someone is infected by bird flu and human flu viruses at the same time then these could easily exchange genes to become more virulent. When not infecting us bacteria (and viruses) can easily be ignored as they are after all rather small and cannot be seen with the naked eye. However, they are so ubiquitous that they make up most of the living mass on the Earth and are likely to be the most common form of life encountered (if any life *is* encountered) on any planet we come across. Still, bacteria and viruses can take part in Darwinian evolution even if they just prefer not to sometimes.

For my own part I prefer a two-stage definition of life.

1) Life utilises energy and materials from the environment to create localised order, i.e. life builds itself and its offspring. This is important when looking for signs of life because, by locally creating order, life affects the environment it occupies. For example plants give out oxygen, which can be detected. Without life planetary atmospheres can contain, at best, very

little oxygen as it is rapidly used up by reactions with minerals in the rock.

2) Living organisms are the result of reproduction and carry genetic information from their parent organisms, which they can, in most cases, transmit to their own offspring.

What about viruses? Some scientists believe that viruses are alive, while others are equally clear that they are not. This puts them close to the border between living and non-living objects, which is where one would hope to find a definition of the simplest life forms.

What are Viruses?

Viruses are very small (about 1/1000 the size of the average bacteria) and just consist of a capsule of DNA or RNA. (DNA is the familiar double helix that carries genetic information; RNA is a simpler molecule, which can, under some circumstances, carry genetic information) and a tail with attachments.

Typically a virus has about 10 genes compared with the 4,500 or so found in common bacteria (even the smallest bacteria contain 500 or so). The capsule does not have a cell wall but a protein coat instead. Viruses also have spikes or legs that enable them to attach to a cell.

Compared to something like a bacterium, viruses are simple but compared with non- living things they are in fact quite complicated. They are obligatory parasites, which just means they cannot reproduce by themselves. They reproduce by attaching themselves to a cell wall and then injecting genetic material (DNA or RNA) into the cell. This genetic material shanghais the normal functions of the cell into producing hundreds of copies of the virus. The cell then dies and bursts open, releasing more virus particles in to the environment.

I have said that viruses are relatively simple, but I would like to qualify that. They are in fact very highly evolved, and can in most cases infect a number of different host organisms. They evolve very rapidly and are incredibly efficient in their use of resources.

It is interesting to compare viruses with other obligatory parasites. Tape worms are well known and widespread; they cannot reproduce outside of the organism they have parasitised and have reduced their bodily functions in such a way that all they really are is a mouth to grip onto the intestines of their host. They do not need stomachs as they are immersed in a nutrient solution and they have a simple form of reproduction (shedding bits of themselves which grow into other tapeworms). So viruses like tapeworms have simply discarded unnecessary baggage in the interest of propagating themselves. Actually it is fairly obvious that viruses cannot have been the first organisms to evolve, as being parasitic they need another organism to be a host.

Of course viruses are not fun when they infect you, but they may very well have been responsible for important evolutionary developments in more complex organisms. Sex is a difficult phenomenon to explain in the light of the obvious fact that reproducing asexually saves enormous amounts of effort. (!!) There are higher animals, for example certain lizards, which reproduce asexually; so one lizard can produce an exact copy of itself (or very nearly, taking into account random mutations) without bothering to find a partner with all that entails. The evolutionary record however shows that, for higher organisms at least, asexually reproducing ones do not last long in evolutionary terms. Sexual reproduction mixes the genes that control the immune system, giving more of a chance of defeating infections - and incidentally driving evolutionary adaptations. Ironically then, viral infections could well be responsible for a lot of evolutionary develop. As a footnote I must mention the recent discovery of an anomalously large virus named Mimivirus that infects amoeba. This virus has 911 genes, more than some bacteria. Seven of these genes are shared with other organisms something not found so far in any other virus. It also contains genes for photosynthesis, genes that should not be present in an obligatory parasite. Not only does Mimivirus contain these genes, which it really shouldn't, but they are apparently active, which begs the question why? Mimivirus is larger than many bacteria but is still a virus, as it does not have

> *a cell wall, although it does have an especially strong protein coat. All this is evidence, as far as I am concerned, for Mimivirus being essentially a missing link between viruses and the rest of life. It simply has not evolved as far as most other viruses (or perhaps the word should be 'devolved', as complexity is being decreased rather than increased, which you would normally expect in evolution).*

You will have gathered that I believe that viruses are alive, but there are cogent arguments that this is not true. The arguments for viruses not being alive are mainly that:

Viruses do not grow and develop; each type of virus is exactly the same in size and construction as any other virus of the same type (barring random mutations and gene-swapping).

Viruses cannot reproduce by themselves. They are purely parasitic needing to infect a cell and use the cells own genetic machinery to reproduce.

Viruses do not have a cell wall.

When a virus is not infecting a host cell it is inert, i.e. it does not react to its environment.

I would argue that although viruses do not develop, once they are in a host cell they do grow, in a restricted manner, in order to reproduce.

Viruses are not the only organisms to rely on other living things to reproduce. Other parasites do the same, but viruses take it to the extreme.

Although viruses do not have a cell wall they do have a protein coat.

Many bacteria are also inert if they are not in the right environment and viruses do react to their environment when they are infecting host cells.

Finally viruses evolve to cope with differing environments, hence different strains of flu etc. No feasible machine is capable of doing this without direct human intervention.

We must now define non-living things, which is hopefully a lot easier as non-living things are inherently less complicated than living things. A chair is usually not alive (you could make a living chair but the living bit would be separate from its chairness) as it does nothing other than fulfil its role as a chair. Aircraft are more complicated in that they move and modern aircraft can even react to their environment without input from a pilot, at least in a limited way. If you had never seen a machine at all you could mistake an aircraft for a bird, but this is only superficial. Aircraft cannot grow and develop, they do not reproduce nor extract energy from their environment, but there is a wild chance that if we did ever encounter an alien machine we might mistake it for a living organism.

Computers are a different proposition, as the hardware (all the physical bits such as processors, power supplies, etc.) has to be considered separately from the software (the programmes that make the computer work). The hardware is relatively straightforward. To reproduce itself physically a computer would have to obtain all the necessary materials and energy from the environment and then manufacture all the parts needed to make a copy of itself. The difference between living and non-living is the difference between growing something and building something. Growing does not require that you have all the bits of a mature entity straight away but building something does.

On the software side there is perhaps a chance of life arising, although it would be limited to living as electronic information inside a computer. Research into artificial intelligence is being undertaken at the present time but has produced nothing that would fool most people into thinking that a programme is alive. Of course there are programmes that mimic some aspects of life, computer viruses being the most obvious. Computer viruses can reproduce (but obviously only within other programmes) however they cannot evolve, as they occupy a very structured environment where even a small change in the virus's programme would cause it to become inoperable.

The definition of life is bound to be somewhat esoteric, but having a clear idea of what is alive and what is not alive is helpful in borderline cases and in unfamiliar environments, such as other planets. However, I think that if we were to find extraterrestrial life it would be obvious.

So, from the above the simplest forms of life would appear to be a virus, as these reproduce and undergo Darwinian Evolution and, despite being rather complex, they are still the simplest system that has these properties. Having given thought to what being alive means it is worthwhile to consider what life is in fact made of, so I move on to the nuts and bolts of organisms.

Chapter 2

What Is Life Made Of?

Almost everyone would agree that the only life forms that we actually know anything about exist on the Earth. As to claims of alien abduction, crop circles, flying saucers etc. I can't help being the teeniest bit cynical. Why would an alien race spend a lot of time and resources going to another planet just to play silly games with people who are not going to be believed anyway?

Life on Earth is basically chemistry, in particular the chemistry of carbon, hydrogen, nitrogen and oxygen. All the other elements that make up life are present in very small quantities, although they take part in essential reactions that keep life going. The human body is about 70% water, which makes you wonder why we aren't a lot sloppier than we are, but then we do have skin to keep the floppy bits together and also bones and muscles to make us less bendy.

The principle reason why water constitutes so much of our body mass is because it is the best solvent around (i.e. it dissolves more substances, more rapidly, than any other solvent). Water is often called the "universal solvent" because of this and is so important because chemical reactions proceed much more efficiently in solution than otherwise, in fact many chemical reactions do not occur at all unless they are in solution.

> ### *Why Water is Such a Good Solvent?*
>
> *First of all water is an extremely strange compound, but it is so ubiquitous that we normally take it for granted. Certainly I don't pour a glass of water and think to myself, "Hey this stuff in my glass has all sorts of weird properties", but it does. H_2O is a very small and light molecule to be liquid at room temperature. Just compare it with carbon dioxide, which is both larger and has a greater mass than water but is quite definitely a gas. Hydrogen has one proton (relatively heavy with a positive charge) in its nucleus, orbited by one electron (much lighter with a negative charge of the same strength as the proton's positive charge). Oxygen has eight protons and eight electrons (also eight or more neutrons in the nucleus, which have almost the same mass as the proton but with no charge). In a water molecule (H_2O) the two oxygen-hydrogen bonds are made by pairing an electron from each atom. As the hydrogen has only one electron, when that is used in the bonding the other side of the atom shows the positive nucleus, unshielded by any negative charge. Water molecules are therefore bipolar (a bit*

> *like a magnet), with the oxygen being slightly negatively charged and the hydrogen being a bit positive. The water molecules tend to stick to each other, the positive and the negative ends attracting each other in a mechanism called hydrogen bonding, and it is this that makes the water liquid at ambient temperatures.*
>
> *Other molecules that are at all polarised are pulled into water by the electromagnetic charges on the water molecules. This is true for the majority of biologically active compounds. On the other hand compounds that are not polar do not interact with water molecules and are therefore not dissolved. This is quite fortunate, as if, for example, the molecules making up our skin were polar we would tend to dissolve whenever we had a bath. (I have to constantly remind myself of this otherwise, being aware that water is the universal solvent, I would avoid bathing, with unfortunate results for my social life).*

Because it is such a good solvent at normally experienced temperatures, and biology in all its many facets relies on reactions taking place in solution, liquid water is essential for life to exist. There are large variations in the amount of water that organisms require and there are those that seem to make do without it but, in all cases, any organism that appears to not require it can be shown to be in a dormant state, i.e. they are not, at this time, carrying out any biological function and are just resting. Many bacteria do this under adverse conditions, effectively hibernating until they encounter liquid water and resume their biological functions.

Water also reacts with a variety of biologically important molecules and the hydrated and dehydrated molecules are very different in their chemical and physical properties. Proteins for example have a specific three-dimensional shape that is necessary for their function. This shape is dependent on how much, if any, water is present.

An important property is that ice is less dense than liquid water, because the hydrogen bonds that stick the water molecules together also act to hold them apart when they are cooled to below freezing point. In all other substances the solid phase is denser than the liquid phase, so water is unique in this respect. As ice is less dense than liquid water it floats, thereby acting as an insulating layer and stopping the underlying water from freezing. The bottom of the ocean is considered to be one of the most likely places for life to have originated, which would not be possible if ice did sink.

Another aspect of water is that it takes a relatively large amount of energy to heat it up and water cools very slowly. In effect water acts as a temperature buffer and it is this "thermal inertia" that has kept temperatures within a range that is conducive to life for billions of years.

Although water comprises such a large proportion of an organism's mass, chemically speaking carbon is the most important element utilised by life. Carbon chemistry is otherwise known as Organic chemistry, as opposed to reactions not involving carbon, which are termed Inorganic. There are many more organic compounds around than there are inorganic ones.

Life, being complex, requires large molecules of different shapes and sizes and carbon is unique in the huge number of compounds that it can form. The reason for carbon's uniqueness lies in the number of chemical bonds that it can form with other chemicals and also with itself. Each carbon atom has the possibility of joining itself to up to four other atoms. Where it bonds to itself carbon can form long chains, rings and any number of three-dimensional molecules. Other elements attached to these structures change the properties of the molecule depending on what the other elements are and where they are located.

The Nature of Carbon

All atoms are made up of neutrons, protons and electrons. Neutrons and protons have, to a first approximation, the same mass, which is defined as 1 atomic unit, or a.u., though protons carry a positive charge and neutrons have none. These heavy particles reside in the tiny space in the middle of the atom, which is called the nucleus. The electrons have negligible mass but carry a negative charge and circulate around the nucleus. It is how the electrons behave that dominates the chemistry. The number of protons in the nucleus of an atom defines what it is (hydrogen has one proton, helium has two, lithium three and so on). The neutrons just add mass, but are of vital importance for holding together the nucleus of any atom heavier than hydrogen. (The forces between two positive charges in such a small space would otherwise be catastrophic.) Atoms of the same element can contain different numbers of neutrons and these are different isotopes of the element. Different isotopes react chemically in exactly the same way, though the heavier isotope sometimes reacts a little slower.

Carbon has six protons and the most common form has six neutrons. This is called carbon-12. Carbon-13 and carbon-14 also exist, with seven and eight neutrons respectively. In biological reactions carbon-12 tends to be absorbed preferentially, as it is lighter. Because carbon has six protons the neutral atom must also have six electrons. The simplest way to visualise the electrons within an atom is occupying discrete orbits around the tiny, heavy nucleus. The first orbit can only hold two electrons, the next one out can hold eight. Carbon therefore has two electrons in its inner orbit and four in the outer one, leaving space for four more electrons from outside the carbon atom. Atoms can form bonds with other atoms by sharing electrons, if there is space, so carbon can form a total of four bonds and indeed it must, as it is an atomic imperative to achieve a full outer shell. These bonds may be with four different atoms, but carbon, being quite small, can also form double or even triple bonds by sharing two or three electrons with just one other atom. This gives carbon an enormous flexibility in its reactions.

The closest element chemically to carbon is silicon, which also has four electrons in its outer shell and regularly forms four bonds. However, silicon is a larger atom, with another complete electron shell between the outer, 'active' electrons and the nucleus. It just cannot snuggle up quite as close to other atoms as carbon, and its bonds are therefore weaker and easier to break. Double and triple bonds are so weak as to be virtually nonexistent.

The simplest situation is where carbon bonds just with itself, but even here the nature of the bonding can cause huge differences in properties.

Graphite is known to anyone who has used a pencil; it is very soft, as it is made up of sheets comprising six carbon atoms tightly bonded together to form a honeycomb structure in a plane one atom thick. Each sheet is loosely attracted to other sheets allowing them to slide past each other easily.

Diamond is not so common (though less uncommon than the major diamond producer, De Beers, would have us think). In diamond the carbon-carbon bonds are all identical and the carbon atoms are arranged in a regular tetrahedral array. It is transparent when pure and, because all the carbon atoms are bonded to four other carbon atoms using up all the available bonds in a rigid three-dimensional structure, it is very hard, in fact the hardest naturally occurring mineral on the Earth. You can demonstrate that these forms are made up only of carbon by burning them in air, the only substance given off will be carbon dioxide (but do get permission from your spouse before setting fire to the wedding ring).

Other carbon compounds that have the same chemical formula but differ in the arrangements of atoms in their molecules can also differ widely in their properties. For example there are three kinds of sugar with the same chemical formula, containing six carbon atoms, twelve hydrogen atoms and six oxygen atoms ($C_6H_{12}O_6$ is the shorthand chemical formula). These sugars are glucose, fructose and galactose. Glucose is the sugar that we have as an energy supply in our blood, fructose is the sugar in honey and galactose is the sugar in milk. Each of these sugars is based on a ring of six carbon atoms with the same number of hydrogen and oxygen atoms attached, but in different places on the ring and this gives them their distinct properties. One property that varies is their taste; although they all taste sweet each of them has a distinctly different sweet flavour.

If you can get such differences from simply changing how bits of molecules are put together, just imagine the different properties you can get by changing the types of atoms and groups of atoms altogether. For instance, adding an oxygen atom to an organic (i.e. carbon containing) molecule. Oxygen is very reactive, so readily form bonds with other compounds. Ethane is a simple organic compound made up of two bonded carbon atoms, each with three attached hydrogen atoms. It is a fairly unreactive gas at normal temperatures and pressures and is insoluble in water. If you add an oxygen atom to ethane (between one of the carbon atoms and one of the hydrogen atoms) you end up with ethanol, the common alcohol that we (or many of us) drink sometimes. Ethanol is a liquid, which is highly soluble in water and very different in nature to ethane. And this was from adding just one atom to a simple molecule. Most organic molecules are much larger and more complicated than ethane and ethanol. With large organic molecules the opportunity for substitution of the active parts is much greater giving a huge reservoir of compounds with different properties, exactly what is needed for life.

Although biological compounds are generally large and complicated they are made up from simpler compounds, hence the sugars above can link together to form long molecules called polysaccharides. The simplest and best known way in which larger molecules are produced from simpler ones is by photosynthesis. Here plants utilise the energy of sunlight to convert water and carbon dioxide into carbohydrates and oxygen (carbohydrates are carbons strung together with hydroxyl groups attached (hydroxyl groups are simply the OH bits from the water)). Carbohydrates can be very long molecules and also can be very rigid, one of their properties that enables them to act as a support for living organisms. It is an interesting fact that the bulk of trees is made up of cellulose, a carbohydrate, which is derived from atmospheric carbon dioxide so trees are literally grown from thin air.

Some of the most complicated and arguably the most important carbon based molecules for life are proteins. These are long strings of amino acids (just another sort of organic molecule with an acidic bit and a basic bit – see box) and are the building blocks of life. For example, some proteins form structures such as hair, feathers, claws etc. Others are catalysts (catalysts help reactions along but do not themselves get used up) called enzymes, which regulate many bodily functions, such as digestion.

Acids and Bases

One very important way to categorise chemical compounds is in terms of their acidity or basicity. One of the simplest ways of looking at this is whether hydrogen ions are lost or gained during reaction. A hydrogen ion is just a hydrogen atom without its electron, so is (usually) just a proton. An acid is a compound that readily loses a proton, sometimes quite forcefully, as is the case for concentrated sulphuric acid, and a base is one that will happily pick up a proton. Again, some compounds, such as concentrated caustic soda, can be quite brutal about snatching protons from wherever they can find them, which is why they are corrosive. One of the most common types of reaction around is neutralisation, where an acid reacts with a base by giving it the proton it so desperately craves.

Amino acids are a special type of organic compound, having both an acidic part and a basic part. The acid end can easily react with the basic part of another amino acid molecule, which is how they build themselves into proteins. Most chemical reactions, and especially biological ones, take place in solution in water. Water itself (H_2O) can split up into a proton, H^+, and a hydroxyl ion, OH^-, which is a proton acceptor. It does this normally, but only to a tiny extent (one molecule in ten million). This means that there is always a small reservoir of protons and proton acceptors, which helps facilitate acid-base reactions and is another reason why water is so essential for life.

As macromolecules go you can't get more complex than DNA. DNA is fundamental to the process of life as it is a molecule that can form copies of itself, which is what organisms need to do in order to reproduce and grow.

> ### *What is DNA?*
> *DNA is a molecule that most people have heard of and know has something to do with reproduction. DNA's overall shape as a double helix is so iconic that it sticks in the memory. If I had to think of what shape to give to the molecule that is most central to the existence and propagation of life I would certainly not have chosen a double helix, the problem with a double helix is that it is too neat, elegant and artistic to be part of life which is generally messy. Whatever, I didn't invent it, so don't blame me for its being unbelievable, just accept that's how DNA is. DNA's job is to carry the code that allows living organisms to reproduce themselves. Fundamental to the DNA code are four chemicals known as bases. They are called adenine, guanine, cytosine and thymine and are generally just referred to by their first letters, A, G, C and T respectively. In DNA these bases are joined together in pairs, by weak hydrogen bonds (similar to the hydrogen bonds in water). Importantly adenine will only pair with thymine and guanine is always paired with cytosine. These pairs of bases form a double helical structure which is held together by sugar molecules, which are themselves held in*

> *place, on the outside of the helices, by phosphates (chemical groups containing phosphorous and oxygen). Each base/sugar/phosphate molecule is called a nucleotide and the four nucleotides only differ in their base, as the sugars and phosphates are identical in each molecule.*

In replication, enzymes unzip the double helix of DNA between the bases, which as I said before are joined together by weak hydrogen bonds. On the two halves of the helix the bases then pick up nucleotides from the surroundings, the inside of a cell. As A only joins with T and G only joins with C the two halves of the unzipped helix complement each other so, when the two halves have picked up the complete number of nucleotides, two identical strands of DNA are produced, which are identical to the parent DNA. Very occasionally a fault occurs in the copying and this creates a mutation.

Apart from replicating itself DNA also controls the production of proteins, which are the basic building blocks of life. Sections of the sequence of bases on each side of the DNA double helix are called genes, so a sequence AATG is a gene, as is GCTAGC. Genes vary considerably in length and in fact these two are really too short, as commonly genes have hundreds of bases. In effect the nucleotides act as letters in an alphabet that define proteins.

DNA is not floating freely in living organisms by itself, but instead it is packed within a chromosome. This packing is quite remarkable as, in humans, a strand of DNA, in an average chromosome, is about four centimetres long and contains about three thousand million bases, all in a chromosome that cannot be seen by the naked eye. The number of chromosomes is different for different species. Normal humans for example have forty-six chromosomes arranged in pairs, one of each pair being inherited from one parent, the other one from the other parent.

Each of these pairs of chromosomes has different functions and of course different DNA sequences. In humans, for example, chromosome 23 (an arbitrary assignment of number but you have to denote the specific chromosome somehow) decides the sex of the individual; two X chromosomes means you are female, an X and a Y chromosome means you are male (for the macho male out there, Y chromosomes are titchy compared to X chromosomes and have a much smaller number of genes). This leads to an equal probability of each sex as mating, to produce offspring, involves one male and one female, and one sex chromosome is contributed by each sex. The possibilities are one X from the female and one X from the male to produce XX which is a girl; the second X from the female and the X from the male also leads to XX and a girl; the first X from the female and the Y from the male leads to XY which is a boy and finally the second X from the female and the Y from the male also produces XY which is a boy. There are sometimes mistakes in copying which can lead to extra Xs or Ys. Some of these drastically affect the child while others seem to have little effect.

With a few exceptions (blood cells, sperm and eggs) all cells in the human body contain the same chromosomes. The cells differ because only some of the genes are activated in each cell and this is the result of the cell's environment and development. A whole plethora of proteins and other chemicals dictate that a muscle cell will be a muscle cell etc. Also some genes are activated in response to outside influences, such as the genes for producing certain antibodies.

Other organisms do not have the same chromosomal make-up as humans, and sex can be determined by more complicated routes, leading to a balance of sexes that is not evenly split. This occurs in social insects like bees. External influences can affect the sex of certain organisms and some cichlids (an African lake dwelling fish) can change their sex in response to the environment, particularly the sex of the other cichlids that they encounter. In many reptiles such as crocodiles the temperature at which the eggs were incubated determines the sex of the hatchlings. The environment can also determine the expression of genes in areas other than sex.

The number of genes does not dictate the complexity of the organism and many plants and protozoa have more than ten times the amount of genes that humans have, but are demonstrably less complicated organisms. Overall however, DNA is the overwhelming influence that determines what an organism is.

Can life exist without carbon and/or liquid water?

Science fiction is fertile ground when it comes to looking for ideas for alien organisms that are not based on carbon compounds. The classic "Star Trek" phrase, "Yes Jim it's life, but not as we know it", adds to the charm of the series. (Although I am concerned about the communications within Starfleet and the memories of the Enterprise crew, as they so often encountered life that was "not as we know it", that you would have thought that someone would start keeping a record, or at least remember what happened during the previous few episodes.) There are a lot of other science fiction writers who have postulated non carbon-based life, but the only justifiable option comes down to life based on silicon. Silicon based life has something going for it as silicon is chemically the closest element to carbon.

Silicon, like carbon, can form four bonds and also bond with itself. However, principally due to silicon being a larger atom, silicon-silicon bonds are weaker than carbon-carbon bonds. The longest chain of silicon atoms that can be reliably formed contains about six atoms, whereas there is no known limit to the length of a chain of carbon atoms. This severely limits the size of any silicon compound and its complexity. Complexity in silicon compounds is also less than in carbon compounds, as silicon forms weaker bonds with many other elements than does carbon. Silicon compounds are also, in general, less reactive than carbon compounds. The common oxide of carbon is carbon dioxide (CO_2), a gas that dissolves in water, opening up the possibility of chemical reaction, and can build long chain molecules through photosynthesis. The common oxide of silicon is quartz (SiO_2), which is basically sand, a solid that is not very reactive and does not dissolve readily in water. If it did then building sandcastles on a beach would be impossible and anyway the beach wouldn't exist. The lack of solubility of silicon compounds in water is a general property.

Although silicon is very abundant on the Earth and in the Solar System, carbon is about ten times more common. This is likely to be true throughout the universe, bearing in mind that heavier elements are produced within stars over a very long time period.

That water is essential for life doesn't really need re-emphasising. There are simply no other liquids that get anywhere near water's ability to dissolve and react with other compounds. Ammonia (NH_3 where N is nitrogen and H is hydrogen) is perhaps the closest, as it is a polar molecule like water and even has some hydrogen bonding. However, it is only liquid at a much lower temperature, so any chemical reactions would be much slower. It is also much more basic than water and this dominates its reactions. Water, on the other hand, can act as both an acid and a base, depending on its environment, and this gives it much more flexibility in its behaviour.

Life based on plasma would be quite spectacular, as it would glow brightly and Startrek has had some rather ornamental plasma intelligences. However, it dissipates rapidly in the absence of containment and strong magnetic fields and only gets its shape from these magnetic fields. The Sun is in fact mainly plasma, which is contained by the Sun's gravity and powerful magnetic field. On the surface of the Sun plasma forms loops and other structures that look fairly complicated, but this complication is due primarily to the Sun's magnetic field and if you look in detail such artefacts are not anywhere near as complex as they seem. Apart from the lack of complexity, an organism made of plasma would find it very difficult to react with the outside world. Plasma is inherently transitory, a plasma loop on the surface of the Sun may seem to hold form for a reasonable length of time but the particles forming it are constantly being replaced, so not much chance for memory to exist.

Another favourite of science fiction writers is the chlorine breather. But if an alien were to breathe chlorine it must be to metabolise it, in much the way we use oxygen. However, chemically chlorine can only form one bond, in contrast to the two allowed by oxygen, which severely limits the kind of compounds it can form. And it's not for nothing we use chlorine as a germicide in water purification; it is inimical to simple life.

The origin of life

The Earth is fairly reliably estimated to be 4.6 thousand million years old (compared to the age of the Universe estimated at about 14 thousand million years old). The surface of the planet has changed quite a bit since those early days of course and the very earliest rocks that have been found (at Isua in Greenland) date from about 3.8 thousand million years ago and seem to have been deposited in water. It is a vain hope to find fossils in such ancient rocks, and anyway very simple early life would be unlikely to leave fossils as such, so scientists have to look for secondary signs that could point to life's existence. Part of the evidence for life relies on the ratio of carbon-13 to carbon-12, as living organisms preferentially accumulate carbon-12 as opposed to carbon-13 in their bodies. The chemical information in these ancient Greenland rocks indicates strongly that life existed at this time.

This does little to pinpoint the origin of life, as there is an absence of evidence for or against life over the first 8 hundred million years of the Earth's existence, although this doesn't mean evidence of absence. What it does imply is that life definitely arose before 3.8 thousand million years ago. Although Darwin himself said that evolution did not shed any light on how life began, he did not have access to modern techniques that at least give strong clues to these origins. One very useful technique is cladistics, which uses similarities and differences to construct a tree of life.

> ***Cladistics***
>
> *Cladistics originally arose from observations of physical similarities and differences between species. For example you can look at the wing of a bird, the wing of a moth and a human arm and compare them. Superficially the bird and the moth may seem to be more closely related to each other than either is to a human, because they both have wings. However, the construction of a bird's wing is very different to a moth's wing, as a moth's wing does not contain any bones and is structurally quite distinct. The bird's wing contains bones, as does the human arm; moreover the bones in the wing and the human arm actually correspond in that they have the same number of bones and they are arranged in a similar fashion to each other. It is clear then that birds and humans are likely to have shared a*

common ancestor, although at some distance in the past. The reason that we can say that this was some distance in the past is that there are other physical characteristics not shared by humans and birds. Feathers in birds and hair in humans for instance, although feathers and hair are both derived from scales, so at some point in the past birds and humans would have shared a common ancestor. Of course we also share a common ancestor with moths, but very much further back.

Feathers in birds are very interesting from an evolutionary viewpoint. It used to be thought that feathers belonged solely to birds, but now it is known that many, usually small, predatory dinosaurs also had feathers. It is always possible that feathers arose independently in both dinosaurs and birds, as there are examples of physical features of organisms that have evolved in parallel in different species and not from a shared ancestor. Eyes for instance seem to have evolved independently at least eight different times. However, there are a lot of other features that birds and these feathered dinosaurs have in common, as well as some that they do not. Where a number of features are held in common by different organisms it is reasonable to see them as being close to each other in an evolutionary sense. Dinosaurs themselves are not really that exclusive a group, as there were various lineages that were not that closely related. Velociraptors and brontosaurs were actually not very similar.

Returning to birds and feathered dinosaurs it is clear that technically the feathered dinosaurs should be called birds, even though they had features that they did not share with birds, or birds should be called dinosaurs. This is my preferred approach, purely because I like the idea of feeding dinosaurs in my back garden.

Reptiles are another case worth looking at. Detailed anatomical comparison has shown that in fact various groups of reptiles are not close relatives at all, so that to call something a reptile doesn't actually mean a lot. Really where you draw the line is subjective.

Of much greater use is simply to look at how closely related different organisms are. A lot of work has gone into constructing lines of descent for life based on physical comparisons, which is all we have for extinct species. This has limitations as obviously only organisms that fossilise will show up in the fossil record and there are modern species that look alike but are genetically distinct. Some species of mussels, for example, cannot be told apart by examination of their visible characteristics, but they cannot interbreed and are genetically very different. In practice cladistics using morphology (the structure and form of organisms) can only be applied, with any confidence, to multi-cellular organisms, whereas, of necessity, the very earliest life will have been much simpler.

More recently cladistics has looked at genes and how they have evolved. This is more likely to be valid, as not all important

> *evolutionary changes will appear as physical differences. Another major advantage is that those genes not strongly affected by the environment tend to mutate at a regular rate, meaning that it is possible to put an approximate relative time scale on the evolutionary tree (bush really). Ultimately it is possible to look at groups of organisms, work out what genes they shared with their oldest common ancestor and thus work out the oldest common ancestor for all organisms. Using genes does have its own problems, as it is only possible to study the genetic make up of organisms presently alive. However, evolution is to a large degree conservative so early genes are frequently carried down to the next generation.*

Using this technique, the oldest common ancestors of all living organisms have been shown to be bacteria that require high temperatures and metabolise sulphur compounds for their energy needs. Nowadays similar bacteria are found around deep-sea hydrothermal vents. Hydrothermal vents are mostly found on mid-ocean ridges, where magma is erupting onto the ocean floor. This heat causes the rocks to crack, allowing seawater to circulate, heating up and dissolving many minerals, including sulphur compounds. Because of the high pressures on the seabed this water can reach temperatures of about $400°C$ without boiling.

So it would seem that life may have originated in these extreme conditions. We should be cautious though, as we can only know the oldest common ancestor. It is quite possible that life originated elsewhere and some evolved to live around hydrothermal vents. This just might have survived when a catastrophe eliminated all other life, leaving only the sulphur and heat loving bacteria to repopulate the Earth. Still, hydrothermal vents must be considered as a likely environment for the first living organisms.

Another way to approach the question of how life originated is to work from the bottom up and consider what conditions could possibly give rise to life. We know that water is essential, so would have to be available for the emergence of life. The oldest known rocks from 3.8 thousand million years ago contain evidence that they were sedimentary, i.e. deposited under water, so clearly water was available at this time.

The period before the oldest known rocks and after the formation of the Earth is known as the Hadean (like Hades) and it is thought that the Earth was a hot and violent place during this period. Whether it was hot or not there are good reasons to suppose that water was present, at least during the end of this period. The origin of this water has often been thought to be impacting comets, although isotopic analysis suggests that a good deal of it came via outgassing from the Earths mantle.

During the Hadean, volcanic activity would have been considerably more vigorous than at present as the young Earth had not lost very much of the heat generated during its formation. This means that hydrothermal vents would not only have been present but would have been much more common than in more recent times.

As well as water there would have had to be organic compounds present and these would have been concentrated at least in some areas. Although life can survive in places where organic compounds are present in quite low concentrations, this would not be a good scenario for life to begin. The precursors of life would have been formed by chemical reactions, which will only work efficiently if the chemicals are concentrated somehow, especially if the reactions are complicated.

The first question must be, "Were organic chemicals present at this time?" There are two ways in which such chemicals could arise on the early Earth: They could have arrived in impacting comets and asteroids, or they could have been generated within the Earth's atmosphere. The first almost certainly happened, as organic chemicals arrive this way today. But although many fairly complex organic molecules are present in comets and asteroids, it is not clear whether there would have been enough of the necessary chemicals in the right concentrations to kick start life.

The idea of synthesis in the Earth's atmosphere has been the subject of various experiments, starting with Harold Urey's in the 1950s. The early Earth is thought to have been a rather stormy place, with lots of lightning. To mimic this experimenters passed an electric spark through a flask containing gases that were thought to be present in the Hadean atmosphere. These experiments did in fact produce large amounts of the organic chemicals that are necessary for life, particularly when the "atmosphere" contained hydrogen. Subsequently such experiments were rather dismissed, as modelling of the Earth's early atmosphere suggested that hydrogen would not have been present. However very recent research indicates that hydrogen *would* have been present and in sufficient quantity to produce the necessary organic chemicals.

Whatever the origin of pre-life organic matter, and I believe both sources almost certainly played their part, the chemicals would need to be concentrated somehow. The biggest player in the field is the idea that they were concentrated on the surface of minerals. Clay is quite a popular choice but modern clays are formed partly by bacterial action, so this may not have been a good route in the beginning, in the absence of bacteria. Back then to hydrothermal vents, which are my present favourite candidates. Around hydrothermal vents iron pyrite (fool's gold) is commonly deposited. This material has a reactive surface, which does concentrate some organic compounds, so is a good candidate, which looks even better when the oldest common ancestor is considered.

There are a number of other scenarios, which cannot be dismissed offhand, and the devil tends to be in the details. One of these details is how did cell walls originate? There are lots of candidates as many organic chemicals have a water loving end (hydrophilic) and a water hating end (hydrophobic) which form layers and cell like structures in water. The hydrophilic end sticks into the water, while the hydrophobic end sticks out, creating membranes that are sheet like or even spherical with enclosed water (plus any other chemicals present in the water). I just wouldn't get hung up on the details, as the first life may not have been cellular anyway. What is clear is that there were plenty of opportunities for life to begin with lots of water and organic compounds. It has often been argued that there was not enough time for life to appear during this era, but the only thing I would say to this is, "What is a long enough time if eight hundred million years is too short"?

In summary I would say that water and organic compounds are so commonplace throughout the universe that given the right environment life is bound to arise. The necessary conditions are just that the water is present as a liquid, there is a means to concentrate the organic matter and enough time is available; so if we look to where the conditions are, or have been, suitable we should inevitably find traces of life. This is a good basic start, but what is still left to be determined is where such circumstances might be found, now or in the past. The next consideration has to be that even if life can arise, can it persist long enough to evolve into intelligent organisms? The nearest possible habitats for life lie with the planets and moons of the Solar System, which is where we go next on a short descriptive tour.

Chapter 3

The Solar System

The easiest and most convenient place to look for life outside the Earth is the Solar System, as it contains the Sun, planets and other objects that we can observe closely and in some cases actually land on. Of course, when we talk about the Solar System, what we really mean is the *known* Solar System. This is sensibly defined as the Sun plus objects bound to it gravitationally. The furthest objects that satisfy this, the long period comets (more of which later), are about fifty thousand times as far away from the Sun as the Earth is. By comparison the furthest accepted planet, Pluto, is only thirty nine and a half times the Earth's distance from the Sun. Beyond Pluto observations are very difficult, so almost anything could be hiding there, giant planets, small black holes or things we haven't yet thought of. You might as well write, "Here there be dragons" as was done by medieval cartographers. On the other hand we don't have to worry too much about such objects, as they would be far too cold for life.

Before looking at the Solar System its origins are important, so a brief history is in order:

The Formation of the Solar System

It is generally recognised that the Solar system is about 4.6 thousand million years old and this has been ascertained primarily by dating of radioactive elements. Initially the material that would go to form the Solar System was a more or less spherical cloud of gas and dust - i.e. a nebula (many such clouds can be observed today and, in quite a few, young stars can be detected). When such a cloud exceeds a certain mass (called the Jean's mass) it will start to contract under gravity. Necessarily such a nebula would rotate as it contracted, simply because any pre-existing rotation will be magnified as the nebula contracted (everyone uses the analogy of an ice skater pulling in her arms which increases the speed of rotation). At a certain time the material in the centre of our nebula became dense and hot enough for nuclear fusion to start, giving us the early Sun. The leftover material remained orbiting the Sun as a disc. Within this disc, by chance, some areas had a higher density than others; the higher density areas attracted more material to themselves and formed small bodies (about 10km in diameter). The largest of these bodies then attracted more material to themselves until there was little material left to grab; at this stage these bodies became the planets. The planets are not in random positions as they interfere with each other gravitationally so there are well-defined areas where planets are stable enough to form.

Because of the closeness of other members of the Solar System and the number of space probes that have studied it we can be certain that the Solar System does not contain intelligent, communicating life apart from the (possible) exception of the Earth.

It is well worth noting that the preceding statement was not true until relatively recently. At the beginning of the twentieth century Percival Lowell observed canals on Mars built by the inhabitants to transfer water from the poles to the equator of the planet. (The "canals" are an optical illusion. Even with a large telescope such structures cannot be resolved when observing from the Earth and anyone who has observed Mars through a reasonably good telescope would testify to this.) Lowell's ideas were not widely accepted amongst astronomers but were believed by many non-scientists. Up until the early 1960s there were still people who would argue for intelligent life on Mars or Venus, but these beliefs were shown to be wishful thinking with the advent of observations by spacecraft.

(The first space probe to send back close up images of Mars was the USA craft called Mariner 4 on the 28th of November 1964. Venus is as yet not imaged very well due to its thick atmosphere, but will be soon by the European Space Agency craft Venus Express. A number of Russian Venera craft have landed on Venus and thoroughly demolished any ideas about intelligent life existing there below Venus' clouds).

Clearly then, within the Solar System we are only looking for simple life forms. However the existence or otherwise of such organisms helps to establish criteria for the habitats needed for life in general.

Having blithely talked about the Solar System a description is in order. The Solar System is about 4.6 thousand million years old (this is the same age as that given for the Earth simply because the formation of the planets took place rather quickly). Physically the Solar System is mostly made up of the Sun, which contains greater than 99.8% of its mass. The Sun lives at the centre of the Solar System and ultimately provides the energy for life on Earth and surface processes on the planets. Other than that it is, at least on first glance, a very average star, unimportant on the cosmic scale - but extremely important on the human scale. It is an incandescent ball of gas with a surface temperature of about 6,000 degrees Kelvin.

*** Temperature Scales ***

Usually in everyday life we use either the Celsius or Fahrenheit temperature scale. Fahrenheit is rather arbitrary so, except in the USA, is not generally used by scientists. The Celsius scale is accepted more widely because it uses 0 °C for the freezing point

> *of water and 100°C for its boiling point (this is at sea level, by definition) and so has reasonable numbers for our environment. The Kelvin (or Absolute Temperature) scale is preferred for astronomy and many other sciences because it has better logic for its use. Absolute zero (or 0°K) represents the point at which thermal movement of atoms and molecules ceases. There can be no lower temperature than this, by definition and it is minus 273°C. An increase of one degree Celsius is exactly the same as an increase of one degree Kelvin. When dealing with a wide range of temperatures having a logical zero point is obviously better.*

I will go into the Sun in more detail later when comparing it to other stars but at the moment the other 0.2% of the Solar System, where life may possibly lurk, has to be looked at first. Distance from the Sun is a major factor in the composition, activity and habitability of the rest of the bodies in the Solar System, so I will work outwards from the Sun in considering these bodies. As measurements that we commonly use are a bit cumbersome on the scale of the Solar System I will use the astronomical unit (AU) to give a better feeling for the distances involved.

> **The Astronomical Unit**
>
> *The astronomical unit is very simply defined as the mean (average) distance of the Earth from the Sun. This is about one hundred and fifty million kilometres. It may seem a bit odd to invent a unit of measurement that doesn't fit in very well with other types of measurement but there are very good practical reasons for this. For a start it is much easier to visualise relative distances within the Solar System, but also it simplifies a lot of the calculations used in measuring distances to other stars.*

The Solar System is a very busy place at the moment so anything I write today could be out of date by tomorrow. The main reason for this busyness is a huge upturn in countries and coalitions of countries getting involved in space exploration. This is brilliant and the information coming in is so very detailed. During the cold war the only players in the game were the USA and the USSR who, for rather silly political reasons, competed with each other to be the first at everything, which often involved cutting corners. Not only did this put lives at risk but also meant that the science involved was sidetracked somewhat. Although the International Space Station has turned out to be a bit of a flop, not having carried out most of its scientific objectives, it was part of a beginning that is now leading to greater co-operation. Now Japan, the European Union, India and China have all joined the USA and Russia in what is turning from competition to closer co-operation, I would predict that many other countries will also get involved in the fullness of time. Given this new impetus I look forward to loads of new discoveries, which will change for the better, our understanding of the Solar System and beyond.

Structurally, the Solar System has most of its matter that is not in the Sun distributed in a rotating flat disc, which is relatively thin. The plane of this disc is known as the ecliptic and is defined, for convenience sake, as being the same as the plane of the Earth's orbit. As well as being in more or less the same plane the orbits of the bodies in the Solar System are overwhelmingly in the same direction and the majority of planets have axes of rotation in the same direction as that of the Earth. These two properties strongly hint that the objects in the Solar System have a common origin, as opposed to being dragged in by the Sun on an ad hoc basis. Bodies that are not actually on the ecliptic (Pluto being the most notable body) are said to be inclined to the ecliptic.

The Solar System can be broken down into different groups of similar bodies at different distances from the Sun. All astronomers will not necessarily accept the criterion I am using, but to hell with it - I think it does at least have some elegance.

Closest in to the Sun are the rocky planets (many people use Terrestrial planets for these objects but, as they are quite distinct from the Earth in many ways, and terrestrial means like the Earth, I prefer rocky planets as that describes quite accurately what they are).

The innermost rocky planet is Mercury. Mercury orbits the closest of any planet to the Sun at 0.39AU, which means that area of the Sun as seen from Mercury is about seven times as big as the Sun appears from Earth. The mass of Mercury is 0.055 that of the Earth, with a radius of 2440km. (As most of the planets in the Solar System are not quite spherical, being somewhat flattened due to spin, the given radius is the average of the two. This makes little difference, as the deviation from a spherical body is usually rather small). Mercury has an orbital period (the length of time it takes to go once around the Sun) of 88 days (in general the closer an orbiting object is to the Sun the shorter its orbital period) and a rotation period of 58.6 days. So there are just three Mercury days in two Mercury years!

Venus is the next of the rocky planets and orbits at about 0.72AU. It has a mass of 0.82 that of the Earth with a radius of 6052km. Venus has an orbital period around the Sun of 224.7 days and a rotation period of 243 days, making a Venus day longer than a Venus year. As Venus is further from the Sun than Mercury, but closer than the Earth, the Sun appears to be nearly twice as big as we are used to.

The Earth has a radius of 6371km and orbits, obviously, at 1AU from the Sun with a period of 365 days and rotational period of 1 day (actually, to be specific, 0.997 sidereal days, which are measured against the background of distant stars and so more precise than our local days.) Sensibly, the Earth-Moon system should be seen as a binary planet, as the Moon has many similarities to the other rocky planets. Its mass is 0.012 that of the Earth, the orbital period about the Sun is obviously the same as the Earths and the Moon has a radius of 1738km. Its orbital period about the Earth is the same as its rotational period (27.3 days) so it always keeps the same face towards the Earth. This is common among satellites of planets and is called synchronous rotation.

Synchronous Rotation

When a satellite orbits a planet its gravity attracts the part of the planet closest to it, causing tides on the planet (tides are also caused by the gravitational attraction of the Sun). These are familiar on the Earth to anyone who has seen the sea (tides on the Earth are not quite this simple, because a major component of them is generated by the centrifugal force that arises because a large moon causes the Earth and the Moon to rotate about a point within the Earth, this is complex and will have to wait for another book). Tides are very obvious in the liquid (and gas) of a planet but also have an effect in the solid part (particularly the squidgy bit beneath the crust).

The energy to raise tides has to come from somewhere, as energy cannot be lost or gained only transformed from one manifestation to another.

With the Moon and tides the gravitational energy of the Moon is being used to produce potential energy in the tide. (Potential energy is essentially the energy needed to raise a mass to a certain height in a gravitational field.) The Earth also drags on the Moon and over the millennia this has caused its rotation to slow until it reached an energy minimum. The minimum energy for the moon is when it has no rotation with respect to the Earth, so the same face is always turned towards us. This is synchronous rotation. As the tidal effect is out of step with the Moon in its orbit (because in the time taken to raise a tide the Moon has moved a little further on) this results in the Moon moving away from the Earth (at present at about 3.8cm a year) gaining potential energy. The Moon is also slowed in its orbit very slightly.

The last of the rocky planets is Mars, which orbits at 1.52AU from the Sun and has a mass of 0.107 that of the Earth and a radius of 3390km. Mars has an orbital period around the Sun of 686.5 days and a rotational period of 1.03 days, very similar to Earth's. Viewed from Mars the Sun has an area of just under a half of that seen from Earth. Mars also has two small moons Phobos and Deimos, which are probably captured asteroids and do not really have much effect on the planet.

Further out from the rocky planets is the asteroid belt. Asteroids are a collection of bodies that are a lot smaller than the planets. They vary considerably in their mass and constitution; some asteroids are made up mainly of a nickel/iron alloy, some are mostly rocky and others have a largely water ice composition. The total mass of all the asteroids put together is at present only about a thousandth of the mass of the Earth, but earlier on in the history of the Solar System it would have been several Earth masses.

Most asteroids are irregular in shape although some of the largest are essentially spherical. It used to be thought that the asteroids were the remains of a planet that disintegrated but it is now believed that they are remnants from the formation of the Solar System that never managed to make a planet, although some were large enough to become similar to planets in their construction, e.g. having differentiated into a dense core and less dense outer mantle. Several hundred thousand asteroids have been detected most of which are small, the largest being Ceres, which has a radius of 457km.

Asteroids orbit in a very loose area between 2 and 4AU from the Sun but can also be thrown out of orbit to either enter the inner Solar System or leave the Solar System altogether. An asteroid with a radius of about 5km impacting the surface is thought to have caused a mass extinction (involving dinosaurs) on the Earth sixty five million years ago.

> ### *The Extinction of the Dinosaurs*
>
> *At the end of a geological period called the Cretaceous (65 million years ago) a mass extinction occurred. This is famously connected with the demise of the dinosaurs, although they didn't really become extinct as one type of dinosaur merely camouflaged themselves as birds (possibly to stop themselves being noticed). Much more important was the extinction of marine organisms and terrestrial plants, especially in the Northern hemisphere. Many groups of organisms had in fact been declining for a few million years before the end of the Cretaceous, so a combination of factors has to be blamed. What is well documented however is the very rapid loss of species at the end of the Cretaceous. This is called the K/T boundary, which is a very clear change from the Cretaceous to the Tertiary*

> *geological period. K is the international symbol for the Cretaceous and T for the Tertiary. So, like most mass extinctions, there was a lead up to the end with a final coup de grace.*
>
> *The coup de grace in the K/T event was almost certainly an asteroid hit. The evidence for an asteroid strike came initially from a thin layer of the rare metal iridium, which is scarce on Earth but is a relatively common constituent of some asteroids. A good candidate for the crater has been found in the Gulf of Mexico, which dates to the right time period (the location is called Chicxulub just to make it easy to pronounce). Gravity measurements (which are necessary as the crater is very deep and under water) show a crater- like anomaly and shocked quartz crystals, indicative of extremely high temperatures, are also found associated with this location. (This glassy material is produced at temperatures greater than can be reached by volcanoes or any other strictly terrestrial phenomena.)*

This sort of impact is very rare now but would have been much more common early in the Earths history, when asteroids were a lot more common.

Beyond the asteroid belt there are the four gas giants and their attendant moons. The gas giants are so called because they are large and made mostly of gas. The gas is mostly hydrogen and helium, which is hardly surprising as these two elements are the most common in the universe.

Jupiter is the largest planet in the Solar System, it orbits at 5.2 AU from the Sun, has a mass 318 times that of the Earth, a radius of 69,910km, orbits the Sun in 11.86 years and rotates once every 0.412 days. From here the Sun's area is less than 4% of that seen on Earth. More importantly, when considering possible life forms, Jupiter has over 60 moons. Of course the tiny Sun means that light levels would not even be as bright as a moonlit night and the heat would be correspondingly low, so not a good place for summer holidays then. The largest four moons of Jupiter are (in order of size): Ganymede (radius 2634km), Callisto (radius 2403km), Io (radius 1821km) and Europa (radius 1565km). Other moons are bound to be recorded, but none near the size of the largest four. At some point it will be necessary to stop counting as objects down to pebble size could be considered as moons if they are in orbit about Jupiter, this would be very silly though as they could exist in the thousands at least. Jupiter also has a very faint ring system surrounding it but nowhere as splendid as Saturn's

The next gas giant out is Saturn, which is the second largest planet in the Solar System. It orbits at 9.54 AU from the Sun, which makes it nearly twice as far out as Jupiter, has a mass 95.2 times that of the Earth, a radius of 58,230km, orbits the Sun every 29.42 years and rotates once every 0.444 days. The most famous characteristic of the Saturn System is its rings. There are many rings (it is difficult to give a figure as they intertwine, have many sub-rings and can be transient (not lasting long)). The rings are made up of mostly icy materials and dust and are probably the remains of a moon that was broken up by collisions. Saturn has over 30 moons, the four largest being Titan (radius 2575km), Rhea (radius 764k, Iapetus (radius 718km) and Dione (radius 560km). The same caveat about not counting moons around Jupiter holds for Saturn.

Uranus is the third gas giant out from the Sun orbiting at 19.19 AU from the Sun, twice as far again as Saturn and about four times as far as Jupiter. It has a mass of 14.4 Earth masses, a radius of 25,360km, orbits the Sun once every 83.75 years and rotates every 0.718 days. Uranus is distinctly odd as, with respect to the plane of the ecliptic, it is lying on its side. This is likely to be the result of a major collision with another large body sometime in the past. Uranus has over 20 moons, the four largest being Titania (with a radius of 789km), Oberon (radius 761km), Umbriel (radius 585km) and Ariel (radius 579km). Note that none of these comes close to the size of our Moon. Uranus also has a ring system more pronounced than Jupiter's but not as noticeable as Saturn's.

The final gas giant is Neptune orbiting at 30.07 AU from the Sun. It has a mass 17.1 times that of the Earth, a radius of 24,620km, orbits the Sun every 163.7 years and rotates once every 0.671days. Neptune has just over a dozen moons but only three are of any size. The three largest moons are, Triton (radius 1353km), Proteus (radius 209km) and Nereid (radius 170km). Neptune also has a ring system but you have to look hard to see it.

Beyond Neptune lies the tiny, icy planet of Pluto, 39.48 AU from the Sun. [Although astronomical units are convenient to use – they do rather gloss over the sheer size of the Solar System. To give a bit of a feel for it I have calculated that if you were to travel to Pluto at the speed of Concorde (Mach 2 – 688m/s), the journey would take about 265 years.] Pluto has a mass of 0.002 that of the Earth, a radius of 1137km, orbits the Sun once every 248years and rotates once every 6.39 days. In these outer reaches of the Solar System the Sun would not even seem to be a disc, being only six hundred millionths of the size as seen from Earth. Charon, Pluto's first detected moon, is not that much smaller than its planet with a radius of 586km. There are two other moons of Pluto that have been recently discovered but they are much smaller. Pluto was discovered in 1930 (Charon in 1978, the other two moons in 2005).

Up until 1992 Pluto was considered to be the furthest planet in the Solar System, then another planetary-like object was detected in the near vicinity, then more, until we now know of several hundred such objects, some possibly even larger than Pluto. It would be nicely dramatic to say that the discovery of these objects was the result of a brilliant scientific discovery, but this was not the case. Gerald Kuiper in the 1950s had predicted such bodies but the techniques of the time did not allow them to be seen. Only when CCD cameras arrived (commonly called digital cameras) could such faint objects be imaged. These bodies are now called Kuiper Belt objects.

There is little point putting a number on the Kuiper Belt objects as they are being discovered at an ever-increasing rate. Presently the Kuiper belt is considered to be a belt of objects roughly 35 – 50 AU from the Sun. There is a vigorous debate going on as to which, if any, of them can be counted as planets. As far as we know Kuiper Belt objects consist mostly of ice and rocks with some organic material. Some may have differentiated into having an essentially rocky core and an icy shell. It has been observed that they vary a lot in their visual properties but they are so far away that interpretation is rather speculative. The Kuiper Belt has been shown to be the origin of the short period comets.

> ## *What is a Planet?*
>
> *Astronomers are surprisingly passionate about what should or should not be called a planet and many hold firm, uncompromising views for one definition or another. Which just goes to show that astronomers are human and like a good argument as much as anyone else. The intuitive definition, based on experience of bodies that are considered to be incontrovertibly planets, is that a planet is a large spherical object orbiting the Sun. This definition automatically includes all the rocky planets, the gas giants and Pluto (which are the traditional planets) and excludes moons. It also has a basis in physics as an astronomical object has to be above a certain size to be spherical. However it also includes Ceres, which is generally classed as an asteroid and an indeterminate number of Kuiper Belt objects, which many people would consider as being something else other than planets. It should be noted however that objects in the asteroid belt were called minor planets, until this went out of favour and there are many people who would designate at least some of the Kuiper Belt objects as dwarf icy planets.*
>
> *Size is often seen to be a major factor in defining a planet but, other than the fact that being spherical depends to a large extent on size, this is rather arbitrary. The historical viewpoint simply accepts that only objects that up until now have been called planets are planets. This is absolute nonsense as the discovery*

of Uranus, Neptune and Pluto occurred after other bodies in the Solar System had been defined as planets and so should not be called planets at all, not to mention the absurdity of not being able to call large objects orbiting another star, planets. My preferred viewpoint would be to consider the Solar System to be comprised of a number of different objects orbiting the Sun, not all of which can easily be given a particular label, as there is a lot of leeway. For instance it is convenient to consider the Moon as a satellite of the Earth when you consider that it orbits the Earth and is smaller than the Earth. It is also convenient when looking at the evolution of planetary bodies to consider the Moon as a planet (and therefore the Earth Moon system as a binary planet) as the Moon shares a lot of characteristics with well-defined planets. So I do not believe that there is a real conflict as a planet is only a planet when you are considering it in that way, otherwise it could be a satellite, an asteroid or even possibly a comet. After all we do only tend to name things as a convenience. This approach of course leads to there being rather a lot of planets in the Kuiper Belt, if you want to look at it that way. I would suggest that you know that the white fluffy thing in the sky is a cloud simply because that is what we call it, on the other hand someone else might well call it something different and so for them it is not a cloud. But it is the same object no matter what we choose to call it.

> **STOP PRESS**
>
> *August 2006: The International Astronomical Union came up with a definition of a planet. According to the new rules a planet must be in orbit around the sun, be big enough for gravity to squash it into a round ball and also have cleared other things out of the way in its orbital neighbourhood. Only the first eight planets achieve all these criteria, leaving Pluto downgraded to a 'dwarf planet'. For the moment the new grade of dwarf planets includes Pluto, Ceres and another Kuiper Belt object, Xena. This list is expected to increase rapidly as more Kuiper Belt objects are discovered.*

Comets

Comets are transient bodies from our viewpoint in the inner Solar System. Somewhere between 20 and 30 a year can be observed by telescope and occasional ones are visible to the naked eye. Comets appear as a nucleus a few kilometres across which, as they approach the warmer inner Solar System, develop one or more tails (these tails point away from the Sun and have nothing to do with the comet's direction of travel). These tails can be several million kilometres long. The average cometary nucleus is tens of kilometres across (although there is no real reason why they cannot be bigger or smaller). The nucleus is usually described as composed of rock and ice often with an organic component. The ice is mainly water ice but carbon dioxide and other volatile substances are present. Some comets may be relatively solid, some like a pile of rubble and some very porous. The tail and coma (a haze surrounding the comet's nucleus) are composed of gas and dust boiled off from the nucleus.

Historically comets were generally seen as portending disasters. Famously Halley's comet appears on the Bayeux tapestry (made to record and legitimise William the Conqueror's 1066 victory at Hastings), which I suppose could be used in evidence for the malign influence of comets, at least from the point of view of the Anglo Saxons. In fact William had difficulty getting his invasion fleet to sea partly because his own men were afraid that the *comet spelled doom for the invasion of England. Eventually William's men proved to be more afraid of him than the comet (his nickname, William the Bastard, was apt in more than one way).*

The earliest known book, from before 2500BC is "The epic of Gilgamesh", which many cultures including the Egyptians, the Jews and the Persians borrowed from. This epic describes the appearance of a comet as being accompanied by brimstone, fire and flood. A similar description appears in many early texts. At least we are no longer so superstitious, what harm can a comet do? Actually a lot of harm. In 1908 a fragment of a comet entered the Earth's atmosphere above Tunguska in Siberia and exploded above a forest. The blast was heard 1200km away and expeditions to the area found 6,000 square kilometres of forest had been destroyed. Fortunately it was in an area with low population. This was only a fragment of a comet, so the effects of a substantial portion of a comet impacting the Earth over a populated area would be devastating. If a large comet were to break up near the Earth (it would be pulled apart by tidal effects due to the Earth's gravity) separate fragments could impact the land and sea giving fires on the land and creating a tsunami, flooding the coast, if exploding over the sea. The brimstone (sulphur) would be released if a fragment hit a volcanic area, as sulphurous gases are emitted from volcanoes. Gilgamesh could well be describing an actual event, as the Arabian Peninsular,

> *where the epic originates, is near both the sea and a volcanically active region. Much of this epic has been shown to describe historically accurate events, which strongly supports the idea that someone (or most likely several people) wrote down a history that was previously transmitted verbally. The epic uses a poetic form, which is the manner in which much verbal information is passed on, as it is easier to remember.*
>
> *Comets fall into two different populations, with little overlap. Short period comets appear at more or less regular intervals of less than 200 years (the 'more or less' is due to the fact that cometary orbits are affected by the gravity of planets they may come near to). Long period comets have hugely drawn out orbits and for all intents and purposes they do not return. Short period comets are also, in general, found close to the ecliptic whereas long period ones can approach from anywhere.*

In a sense the Kuiper Belt objects can be seen as the limits of the Solar System as they lie more or less on the ecliptic so fit in neatly with the Solar System being a disc of matter orbiting the Sun. The long period comets however need further explanation, as they come from any direction and are not restricted to the ecliptic. In 1950 Jan Oort, a notable Dutch astronomer, analysed the orbits of the known long period comets. None of them could be shown to come from interstellar space but they could come from any direction. What Oort found was that these comets were coming from a region located around 50,000 AU from the Sun (different authorities give different distances but they all agree that long period comets come from tens of thousands of AUs away). Because these comets do not approach from a preferred direction he concluded that they come from a spherical cloud surrounding the Solar System where encounters with other comets caused them to fall inwards. This is now called the Oort Cloud. No object actually within it has yet been imaged, as they are too small and far away, but mathematical analysis of cometary orbits supports the existence of the Oort Cloud.

It took a while but it is now accepted that the comets in the Oort Cloud originated from within the Solar System (oddly, closer to the Sun than the Kuiper belt). These comets are now believed to be bodies that were thrown out of the Solar System in all directions during it's early formation, due to gravitational encounters with other bodies. These bodies would not yet have reached planetary scales. The objects that were thrown out would be slowed by the Sun's gravity until they ended up in a shell far from the Sun, thus forming the Oort Cloud. The number of bodies in this cloud is not known but the total mass is likely to be equivalent to several Jupiters, given that the probability of a perturbed object leaving the Solar System proper is probably greater than it remaining in the disc.

This chapter has been the one with the most numbers in, but you should now have a reasonable 'feel' for the Solar System. It was necessary to put things in context before getting down to habitats for life. The next chapter concentrates on the Earth, using it as a yardstick for habitable environments. I find the Earth fascinating and I hope you do too.

Where Are All The Aliens

Chapter 4

The Earth

I am going to talk about the Earth first, rather than take a more logical tour of the Solar System from the Sun outward. After all we live here and are beginning to understand how it works. We can use knowledge gained here to inform our understanding of the rest of the Solar System.

Scientists have long argued that the Earth contains life and some have even postulated intelligent, communicating life. This hypothesis was tested by the Galileo spacecraft in 1990, which observed the Earth in passing on its way to Jupiter. The Moon was also observed.

The results from Galileo were encouraging - sure signs of life were detected. A near infra-red spectrometer detected ozone (which implies the presence of oxygen) and large amounts of methane. Oxygen and methane react readily with each other and so cannot be present together in the atmosphere unless they are being replaced on a large scale; the only feasible mechanism is life processes. In the radio part of the spectrum Galileo detected a puzzling phenomenon. Right across this part of the spectrum were multiple modulated signals in very narrow wavebands. The conclusion drawn from this was that at least some of the life forms were using the radio part of the spectrum to communicate with each other. This is likely to be technological rather than organic, as radio communication, simply because of its long wavelength, requires large transmitters and receivers, the possession of which would be too costly from an evolutionary viewpoint.

These observations raise the question, "Is there intelligent life on the Earth?" No strong directional signals were detected that were aimed to anywhere outside of the planet, which argues that we are not dealing with an intelligent, communicating life form. However, some signals were strong enough so that, with sufficiently sensitive receivers, they may be detectable outside of the Solar System. It seems that with the Earth we are dealing with organisms that are just about to become intelligent communicators (barring this species' extinction of course). When Galileo looked at the Moon it found no sign of life, probably not a surprise.

In practice of course we do not have to rely on the Galileo data, although if another planet was being investigated this is the sort of data that we would have to rely on. The advantage of looking at the Earth first is that we can observe the planet closely, as after all we are infesting its surface. By now we know a good deal about our planet and our knowledge is increasing all the time. First of all we need to consider the physical Earth and what affects it.

The Internal Structure of the Earth

The Earth is very nearly a sphere with a radius of 6370km. The centre of the Earth (the core) is composed of a solid iron/nickel sphere with a radius of 1,215km. This is surrounded by a liquid outer core comprised of iron with lighter elements (silicon, sulphur, oxygen and magnesium have been suggested); this outer core is 2255km thick. Above this is the lower mantle, which is primarily comprised of silicate minerals and is 2230km thick. Further above this is the upper mantle, which has the same chemical composition as the lower mantle but differs in the minerals present due to the higher pressures lower down. (Minerals can have the same chemical composition but differ in their structure depending on the pressure they form at, which is due to the mass of material pressing down on them and so equates to different depths. The best known example is diamond, which is exactly the same as graphite chemically but is very different structurally, because diamond forms at a higher pressure and temperature than graphite). The top of the upper mantle lies at a depth of about 670km. Above all this is the crust, of which there are two types, oceanic and continental. Oceanic crust is denser than continental crust and its thickness varies from 0 to 11km, whereas continental crust varies from 25

> to 90km. It is not uncommon to lump things together, using mantle to mean both upper and lower mantles and core to mean the inner and outer core. Geologists also use the term lithosphere to mean the crust plus the upper rigid part of the mantle. Although the mantle is solid it can act as a very viscous liquid under the high temperatures and pressures at depth, so it can convect. This means that hot material can move from the depths, carrying heat with it, much more efficiently than conduction, the other main heat transfer mechanism.

What goes on miles beneath our feet may, on first sight, seem irrelevant to life on the surface, but it provides our raw materials, including gases in the atmosphere, as well as contributing to the surface temperature. In fact our surface is constantly being reworked and recycled by Plate Tectonics.

> ### *Plate Tectonics*
> *In the late 1960s it was discovered that the Earth's crust is divided up into a number of plates that are moving relative to each other at different speeds. There are two main ways to demonstrate that the Earth's plates are moving with respect to each other. Hot spots are areas of volcanic activity within plates and are thought to occur when a plume of very hot lava, originating at the core mantle boundary, rises to the bottom of*

the lithosphere and produces very large lava flows on the surface. Hot spots seem to be independent of plate movement as they originate much more deeply in the mantle and, as the plates move over a hot spot, a linear sequence of volcanoes are produced, which can be dated and therefore give a measure of the plate's rate of movement. The most famous example is the Hawaiian chain of volcanic islands, where the movement of the Pacific Plate has carried islands formed from volcanoes north-westward implying a north-westward movement of the Pacific Plate. There are plenty of other examples that imply that hotspots are static with respect to the plates, even remaining stable when crossing plate boundaries. It can be argued that although all the hotspots known are unmoving with respect to each other, they may be moving as a group with respect to the Earth itself, but they can still be used as a valid frame of reference.

The Earth's magnetic field magnetises many rocks when they are still in a semi-liquid state and this remnant magnetism in rocks can be detected. As the Earth's magnetic field reverses its polarity at irregular intervals, this remnant magnetism can be used to plot the movement of the plates and also give an indication of the speed of movement.

As the Earth is not growing in size these plates must interact with each other. At the boundaries of plates one of three things is happening; the plates are moving apart from each other, they are colliding with each other or are sliding past each other. Plates can be oceanic, continental or a mixture. Oceanic plates are denser and less thick than continental plates. At the point where plates are moving apart there isn't just a big hole, instead magma wells up from the Earth's mantle to form a mid ocean ridge. As the two plates move apart new oceanic plate is formed and moves away from the mid ocean ridge. (New oceanic plate is also formed when a continental plate splits and the two bits separate, in this case a new ocean may develop). As the newly created oceanic plate material moves away from the mid oceanic ridge it cools and becomes denser.

Where plates move towards each other what happens depends on the type of plates involved. For two oceanic plates the most dense (also the oldest) slides down underneath the less dense plate in a process called subduction. The subducting part of the plate drags the rest of plate with it as it sinks, including a proportion of the sediment on its surface. This is the main force behind plate movement. Sediment on the subducting plate may be scraped of to form islands. A good part of Japan is formed this way. In some cases a bit of oceanic crust may also be scraped off and end up on land as has happened in the Lizard, in Cornwall.

As the down-going plate gets deeper into the mantle it heats up and eventually partially melts. At the point on the overriding plate that sits above the melting, subducting plate volcanic *eruptions occur. These produce, among other volatiles, carbon dioxide, some of which came from subducted sediments, so this is a way of recycling the carbon dioxide that was entombed in the organic portion of the sediments. As well as recycling carbon dioxide the island arc volcanoes create new continental crust as the lighter fraction in the melt comes to the top. When an oceanic plate meets a continental plate the oceanic plate will naturally slide under the continental one. Volcanoes will be formed at a distance from the subduction zone and this is how the Andes were formed. The Andean volcanoes form a much larger edifice than island arcs, as more continental type material is available for mountain building.*

When two continental plates collide neither will naturally slide under the other. Instead the force of the two plates colliding buckles both of them to form mountains. A lovely example of this is the Himalayas, where the Indian subcontinent is pushing into the Eurasian continent. Inevitably there would have been oceanic plate material between the two continental plates before they collided. This is trapped and uplifted to become part of the mountain belt. There are shellfish fossils on top of Everest, which could explain why so many people want to climb it. The collision of two continental plates cannot go on forever and an adjustment of the plates' direction of movement occurs.

Plates that slide past each other tend to scrape off bits of each other and transport it in their direction of movement, leaving

> *exotic (in this sense exotic means rock that is moved to an area where it would not be expected to have formed) bands of material spread along the sides of each plate. Earthquakes are common along such plate boundaries. The San Andreas fault, running through California, is at the junction of the Pacific Plate, which is moving Northeast, and the North American Plate, which is moving Southwest.*
>
> *One final point is that for subduction to occur the presence of liquid water is necessary. This has serious implications for life trying to establish itself on a dry planet.*

Plate tectonics is involved in the recycling of carbon in several ways. Organic carbon (bits of dead organisms and faecal material) rains down onto the floor of the oceans and forms sediment. It is then carried with the oceanic plate away from the spreading ridge and towards the subduction zone. When it finally reaches the subduction zone much of the organic matter is carried down with the descending plate. Eventually, most of the carbon, in the form of carbon dioxide, is belched out by volcanoes on the overriding plate. On average it takes about two hundred million years, a relatively short time in a geological framework, for organic carbon to be deposited in the oceans and then re-emerge as carbon dioxide via volcanoes.

Sediment that does not get subducted accretes (is scraped off) onto the edge of the overriding plate and adds to the continental crust, being forced upwards in response to the pressure exerted by the subducting plate. Parts of Indonesia were formed this way.

The fate of organic carbon that finds itself incorporated in continental crust is somewhat more erratic than that on ocean crust. Continental crust returns organic carbon to the atmosphere by uplift and erosion. Essentially tectonic activity forces mountains upwards, where they are exposed to wind and weather and so get worn away more rapidly. When rocks containing organic carbon are exposed, the organic matter is washed out, carried away by streams and rivers, and eventually ends up in the oceans. This mainly forms sedimentary rock on the continental shelf, although some is carried further out to sea and ends up on the nearest oceanic plate, where it may be subducted in the fullness of time. The result of all this activity is a constantly changing physical environment.

A very striking aspect of Life on Earth is that there is so much of it and almost wherever you look. No matter what the environment, there are organisms that are adapted to it. (The only place I can think of that does not contain life is molten lava and I may be proved wrong, but I don't think so. Even so, when it cools it is rapidly colonised.) We have recorded over two million species, but know that to be a huge underestimate. Estimates range from five million to one hundred million but even then bacteria and other single celled organisms tend not to be counted, despite representing most of the biomass of the Earth. Come to that, parasites are rarely considered, though most life has its parasites, often multiple ones and often these parasites are species specific (i.e. they only infect one species). I must admit that the slogan, "save our parasites", is unlikely to gather as many adherents to the cause as, "save the dolphins", has but parasites fulfil an essential, if rather distasteful, role in the environment. There is even evidence that, paradoxically, some organisms benefit by hosting parasites - possibly because they stimulate the immune system.

Life changes its environment with feedback mechanisms that help to maintain it. This concept was taken to the extreme by James Lovelock with his concept of Gaia, the Earth as a living planet, keeping its environment stable and suitable for life. It is true that conditions on the Earth have, since life began 3.8 thousand million years ago, always allowed life to exist without total extinction, even though the temperature of the Sun was some 25% – 30% weaker than now. At present we can see feedback mechanisms between life and the planet that help to maintain a suitable environment for life.

Feedback Mechanisms

Feedback mechanisms are either positive or negative. A positive feedback mechanism reinforces a change, whilst a negative feedback opposes it. The climate exhibits some of the most obvious and important feedbacks. An example of a positive feedback is the growth of glaciers. If a glacier expands, because the climate is cooling, then it reflects more sunlight and this leads to further cooling and growth of the glacier. Negative feedbacks also operate in a cooling climate. Colder conditions mean less water is evaporated from the oceans and this means less precipitation, so the growth of glaciers is inhibited. As a result less sunlight is reflected and the cooling is slowed down.

In a warming climate more water is evaporated. Water vapour is a greenhouse gas, so a positive feedback results, increasing the temperature. However, more evaporation means more precipitation. Rain reacts with carbon dioxide to produce carbonic acid, which dissolves rock forming insoluble carbonates. This removes carbon dioxide from the atmosphere, so reducing the greenhouse effect. This is needless to say a negative feedback.

There are many more feedback mechanisms in operation so the situation is very complex, a nightmare for climate modellers. In general, however, these mechanisms usually work to stabilise the climate. Where things go wrong and climate deteriorates is where the rate of change in the environment is rapid, as feedback mechanisms tend to act over a long period of time. Anthropogenic (i.e. due to human activity) release of carbon dioxide into the environment has been very rapid since the

> *industrial revolution and is overwhelming the natural feedback mechanisms, thereby causing global warming. It is not at all clear that a runaway greenhouse effect can be avoided.*

Although feedback mechanisms may stabilise the Earth's environment under some conditions it is clear that this is not always the case. There is strong evidence that the first organisms were anaerobic (they prefer an environment without an appreciable amount of oxygen). A proportion of these anaerobes were photosynthetic, consuming carbon dioxide and releasing oxygen, thus poisoning their own environment. On a static planet life would therefore be marginalised, as local conditions deteriorated. The Earth however is not static, as the surface is constantly being recycled by plate tectonics.

Land plants appeared about four hundred and eighty million years ago. (Remember that life in the oceans started over three thousand eight hundred million years ago.) The very presence of life on land had a major effect in speeding up the carbon cycle.

Think of a tree. First of all its roots physically break up soil and rock, as well as pumping out chemicals to dissolve minerals. As it grows it incorporates carbon from carbon dioxide in the atmosphere into its structure. When it dies the remains may be buried, turning into peat and eventually coal.

More likely though, it will decompose on the ground, releasing its stored carbon over a matter of months, or at most a few years. If it's unlucky, it might catch fire (before humans were around fires were commonly started by lightning) and so release all its carbon very quickly indeed, in just a few hours.

Compare this to algae floating in the sea. These also incorporate carbon from atmospheric carbon dioxide into their structure. However, when they die they fall onto the sea bed (oceanic plate), where they are carried along for millions of years until they are eventually subducted and have a chance of releasing their carbon back into the atmosphere.

One important difference between oceanic and continental plates is that oceanic plates do not last nearly as long. There is always a subduction zone waiting to swallow them, whereas the continents, being lighter, float on top. The continents are therefore a huge reservoir for organic carbon.

The process that incorporates carbon into living matter (photosynthesis) also releases oxygen. Overall carbon dioxide is locked up in the rocks for hundreds of millions years and the oxygen released builds up in the atmosphere. This really is where the Gaia hypothesis falls down most dramatically. In the early history of Earth anaerobic bacteria released oxygen that then built up in the atmosphere and poisoned them. (Nowadays these bacteria occupy extreme environments where there is no oxygen.) This probably represents the first mass extinction, although it is never acknowledged as such. Bacteria are simply not cuddly.

One side effect of this release of oxygen was the build up of the ozone layer that protects the Earth from damaging ultra-violet radiation, which directly harms genes. Ozone is simply three oxygen atoms bonded together which strongly absorb in the ultra-violet part of the spectrum. Ozone is easily broken down so needs a constant supply of oxygen to maintain it. At the moment the ozone layer has holes in it, around the poles, letting in ultra-violet radiation. This is a man made effect, caused by the release of chemicals that help break down ozone.

On the subject of dangerous radiation, the Earth is constantly being bombarded by charged particles from the Sun and cosmic rays from space, these being somewhat nasty things. The Earth's magnetic field diverts this radiation (so of course the surface of a planet without such a magnetic field would be bombarded by it). The Earth's magnetic field reverses, at what seems to be random times. Although the details are very complicated, in essence, during such a reversal the field drops close to zero and should let radiation through. Oddly there is little evidence of enhanced extinctions during these reversals.

So the Earth's system of life and geological processes fails to maintain life in the manner that it would like to be used to. However, the Gaia hypothesis did, at least, call people's attention to the interrelation that goes on between life and Earth. In fact the very instabilities that are inherent in the Earth's biosphere are themselves closely involved in the evolution and radiation of life forms. On one extreme there is the "Snowball Earth" theory.

> ### *The Snowball Earth Theory*
> *There is strong evidence that between 750 million years ago and 580 million years ago there were several (you can find estimates of four to seven in the literature) glaciations that totally covered the Earth from the poles to the equator, each coming to a sudden end. I naturally shy away from exotic scenarios as there is usually a more mundane explanation, so originally considered a Snowball Earth highly unlikely as I believed that once a global freeze occurred, the Earth's surface would reflect back so much solar radiation that a recovery was not possible. I still maintain that there must have been at least small areas not frozen, for example on the flanks of volcanoes. Other than that however the evidence is pretty conclusive that these glaciations took place. The basic evidence is found in rocks of the period, which were on the Equator at this time, but are associated with clear evidence of glaciation. These rocks have iron oxide deposits intermixed with them (which imply periods of very low oxygen levels followed by periods of high levels of oxygen) and are capped by carbonate rocks, which only form in tropical*

> *conditions. This implies a series of extreme glaciations, which would have caused a massive crash in the number of species and overall variety of life.*
>
> *The glaciations then ended, probably because carbon dioxide from volcanoes increased in the atmosphere, building up to a very hot greenhouse effect.*
>
> *The Snowball Earth was probably initiated by a number of factors. At this time virtually all the Earth's continental plates were grouped around the Equator and largely stuck together into a massive plate. There was therefore less plate movement, fewer spreading ridges and subduction zones and consequently less volcanic activity and less carbon dioxide released into the atmosphere. (A decrease in carbon dioxide levels gives global cooling, just as an increase leads to global warming.) Because all the land was in one big lump there was a lot less coastline – and therefore a lot fewer shallow seas – than we have now. As shallow seas provide the most prolific environment for life, there would have been less biogenic carbon dioxide. And finally, the Sun was a bit (about 15%) cooler than the present day.*

The effect of the Snowball Earth would have initially devastated life forms, reducing them to a few refuges (at this time there were only very simple organisms). The sudden end to the glaciation subjected life to hot house conditions in a very short time period. Not only did carbon dioxide build up, but the warming would have released methane, a very potent greenhouse gas. The methane was stored in cage-like molecules called clathrates, organically derived methane locked into water ice, which were quickly oxidised but would have done the damage in the meantime. Methane reacts with oxygen in the atmosphere (helped by ultraviolet light from the sun and/or lightning strike) so much so that the presence of appreciable amounts of oxygen and methane together is an incontrovertible sign of life.

At first glance these dramatic oscillations of climate would seem to be detrimental to life but oddly, during this time, there was an unprecedented blooming of increasingly complex and diverse life forms. Life almost certainly existed 3.8 thousand million years ago and up until the Snowball Earth period was no more complex than simple filamentous algae. Suddenly at about 670 million years ago, during the Snowball Earth period, complex multi-celled organisms appeared.

Why multi-celled organisms didn't appear for over three thousand million years is very difficult to explain. It's very tempting to say that it is intrinsically difficult for such organisms to evolve and this only can occur under very specific conditions but that really doesn't explain anything. The hothouse period after the glaciations saw not only an unprecedented explosion of species, but also a massive increase in biomass (the weight of organisms). This was probably the mechanism that stopped a runaway greenhouse effect as organisms gobbled up the carbon dioxide from the atmosphere.

As a direct comparison to the Snowball Earth there is the much more recent Cretaceous period, which lasted from about 142 million years to 65 million years ago, when a mass extinction occurred famously killing off the dinosaurs. Not quite all of them, as I have intimated before, I enjoy putting feeders in my garden to feed the small, flying feathered dinosaurs that survived.

The Cretaceous Period

The Cretaceous period is notable for being an exceptionally warm period, with temperate forests at both poles and probably the highest sea level that the Earth has experienced. There are three main reasons for this high sea level. Firstly, the higher temperatures would have lead to thermal expansion of the oceans, (which just means that the same amount of water takes up more volume when it's hot). Also there was greater seafloor spreading, pushing the mid-ocean ridges upwards and so displacing vast quantities of water. Last, but not least, there was no permanent ice cover at the poles. Today it has been estimated that 29 million cubic kilometres of water are locked up as glaciers and polar ice. If this all melted it would lead to a

73 metre rise in sea levels, not even considering thermal expansion. In the Cretaceous this displaced water covered the continental shelves and low-lying land. The centre of the North American continent was flooded, as were large areas of what would become Europe.

The high temperatures were due to high carbon dioxide concentration in the atmosphere (ten times that experienced at present.) Higher carbon dioxide levels were the result of the break up of two major continents, Laurasia in the north and Gondwana in the south. This led to a greater length of spreading ridges, with their associated volcanoes, so more carbon dioxide was emitted giving an enhanced greenhouse effect. In addition, there was no ice to reflect sunlight back into space, so more energy from the sun reached the Earth. The ocean currents distributed warmth more efficiently because of the position of the continents at this time.

The Arctic and Antarctica are probably the most unlikely candidates for a forest-based ecology. At this time the northern continents and Antarctica would seem to have been in approximately the same positions as now and so would have experienced months of total darkness, hardly conducive for a forest. This is one of those, "What on Earth are you talking about?" moments, "You simply can't have forests at the poles as it goes against common sense and everything that we are used to". Put into context, the Arctic and the Antarctic had temperatures comparable to the South of England today (average winter temperatures of about 6°C and average summer temperature of about 21°C). But although temperatures may have been comparable, the overall environment was unlike anywhere in the present day, because of the polar light regimes with months of darkness. Most plants got over this by being deciduous and shedding their leaves before winter.

In the Arctic there were large animals, principally dinosaurs, which migrated South in the winter. Good examples are the Hadrosaurs. These were bipedal dinosaurs growing up to 10 metres long and weighing up to 10 tons. They had small front legs and much larger rear legs, with webbed feet, possibly for walking through swampy terrain. Hadrosaurs were principally browsers, feeding largely on shrubs and trees. Interestingly, it used to be thought that grass was very scarce in the Cretaceous, but recent evidence from coprolites (fossilised dinosaur poo) suggests that grass was more widespread than previously thought, so the Hadrosaurs probably also grazed. Their food supply would all but disappear as the trees shed their leaves, hence the need to move South for winter. Like many herbivores, Hadrosaurs travelled in herds for protection from following carnivores. Imagine wildebeest migrations, but instead of small cattle-like animals, a herd of animals seven times as tall and five times as heavy coming towards you at 40 miles an hour (dinosaurs were not as slow as is often thought). A great

> *natural history film.*
>
> *Less is known about Antarctic environments as fossil hunting is strictly limited, but similar constraints would have operated.*
>
> *At lower latitudes, nearer the equator, it was hot and humid with some areas averaging over 40°C (104° Fahrenheit). The most conspicuous natural phenomena were extensive carbonate platforms (areas of chalk built up in the shallow seas by the shells of small planktonic organisms and shelly material from benthic (bottom dwelling) molluscs). The white cliffs of Dover provide a good example from this time. These deposits were world wide and much more extensive than anything like them today. The reason they were so extensive is because of the large area of shallow seas present in the Cretaceous, especially on drowned portions of the continents. Also with the higher temperatures partially enclosed bays would evaporate, encouraging deposition of dissolved carbonates.*

The Cretaceous environment seems unusual from our perspective, but if we take the long view backwards over the last 354 million years, then for 80% of the time the climate was warmer than it is now. It is tempting to look at the high carbon dioxide levels in the Cretaceous as a model for the Earth in the near future and think, "This is not so bad, a nice warm climate and, judging by the dinosaurs, life was pretty good". However, when you consider the drawbacks, life wasn't that rosy.

During times of particularly vigorous volcanic activity the huge amount of carbon dioxide spewed out was accompanied by sulphur dioxide, leading to acid rain (not very good for watering your plants) and a general deterioration of the environment.

Of particular interest are the Deccan Traps that are found today on the Indian sub-continent. The Deccan Traps erupted towards the end of the Cretaceous and are exceptionally large volumes of lava that originated in a mantle plume, that is very hot magma coming from deep in the mantle and carrying material right to the surface. This hot material erupted rather gently, and flooded a vast area over a geologically short period of time, pumping huge amounts of carbon dioxide, sulphur dioxide and other gases into the atmosphere. At the same time similar eruptions were taking place under the oceans, particularly the Pacific and Indian oceans.

Carbonate platforms were a major sink for carbon dioxide but the evidence shows that their existence was episodic, with times when they were very extensive and other times when they were quite sparse. This would have lead to climatic swings, stressing the environment as the carbon dioxide levels fluctuated. It should not be forgotten that the higher sea levels drowned much of what are today heavily populated areas. Actually, this might not be too bad as cities are a major source of carbon dioxide and other greenhouse gasses, so drowning them could be seen as negative feedback in action.

The instability of the environment is apparent in the fossil record, with the Cretaceous having high rates of extinction and speciation - i.e. lots of new species evolving to try and keep up with the changing environment. Towards the end of the Cretaceous a distinct deterioration of the climate was occurring. On land the dinosaurs were in trouble and in the sea ammonites and sea-going reptiles, such as ichthyosaurs and plesiosaurs, were rapidly becoming few and far between. Pterosaurs disappeared altogether.

At the end of the Cretaceous the impact of a largish asteroid, at Chicxulub in Mexico, marked the final extinction of many species. First of all the energy of the impact vaporised much of the target area and sent out a devastating shock wave, felt around the globe. The intense heat caused massive forest fires and dust and debris was thrown miles up into the stratosphere, blocking the sun for several years. Not only that but there would also have been widespread tsunamis. However, from what has been said before, this was probably just the final straw in an ongoing extinction crisis.

In fact the K/T boundary is not as sharp as was once believed. Ammonites, which were thought to have been wiped out by the asteroid impact, have been found in strata 200 thousand years after they were supposed to have become extinct.

Although not fully verified there are fossils of dinosaurs found in South America and China that seem to have lived, at least for a short time, after the impact and don't forget the birds. Following this, at the start of the Tertiary geological period, there was a fairly rapid increase in the number of new species. Mammals in particular were favoured, both on land and in the seas and the surviving dinosaurs, in the guise of birds, were very successful.

The K/T extinction was not unique. There have been mass extinctions several times over the history of life on Earth, probably for a number of different reasons, such as climate change (warming or cooling), changes in ocean currents wiping out habitats, atmospheric pollution and extreme volcanic activity, - as well as possible extra-terrestrial events. Mass extinctions do stand out (they are after all <u>mass</u> extinctions), however, there is always a background level of extinction as species come and go.

Interestingly, species do seem to have a limited length of time before they become extinct and an average of about 0.5 to 1 million years for mammals and birds and 4 to 5 million years for shellfish seems to be reasonable from the evidence. The difference between these two rates may be an artefact of measurement. The most common fossils are shellfish, because they are quite easily fossilised, the sea bed is comparatively stable and at death organisms are often buried fairly rapidly. On the land erosion is much more fierce and only in unusual circumstances are fossils well preserved, so there are many gaps in the fossil record. If the rates do really represent the length of time a species exists it could simply be that there are more varied habitats on land. This would allow lots of different species to develop, but extinction is more likely because terrestrial environments respond to climatic change in a much shorter length of time than under water. It should also be realised that the environment for any species contains other competing organisms (especially of the same species), so a larger number of habitats implies more competition and a rapid turnover of species.

Of course for the number of species to remain constant the rate of new species arising has to equal the number of species going extinct. What actually happens is that the number of new species evolving is somewhat greater (on average) than the number of species going extinct, so overall, the number of species increases over time. This is not an effect of there being a greater chance of finding relatively recent fossils as this has been taken care of in the calculations.

Sometimes extinctions occur at a rate that is faster than the background, but less than in mass extinction events. Local extinction of endemic species (species that occupy a limited habitat) has always been common, as over-specialisation makes species more sensitive to even small changes in their environment.

The story that comes out of this is far from the Gaia concept, or the implied smooth speciation of early evolutionary scenarios. What the evidence is saying is much more dynamic. Even over time periods as short as a human life-time the climate is not stable and niches come and go. Organisms adapt or die out. Although evolution is still in operation, such short periods seldom produce new species, although surprisingly, it does occasionally happen. Rather, new <u>varieties</u> arise that may be on the way to becoming species but can still interbreed.

It is almost always considered bad science to even think that we are living in an environment that is unique in any way. However, because the environment is continually changing the Earth can be seen as a whole set of unique situations. For instance, we are living in the first ice age for 270 million years. This ice age started just 1.8 million years ago, a mere moment, geologically speaking, and has seen the emergence of humans. We have quite a lot of information about this era, simply because it is so recent. The period since the beginning of the ice age is called the Quaternary.

The Quaternary Age

After the Cretaceous Age ended 65 million years ago there was a gradual cooling of the climate until we get to 1.8 million years ago, when substantial ice sheets existed at the poles. Since then the ice has advanced and retreated several times, but has never completely gone away. Drilling into the ice sheets, along with cores drilled into the ocean floor, gives us a detailed picture of the temperature swings that have occurred since then. These cores show dramatic rises and falls of temperature, sometimes happening in as little as a few decades or as much as the odd thousand years or so. Certainly the recent fossil record shows that species can rise and fall in periods as short as a few hundred years. Information from before the ice age seems to show changes happening more gradually, but we have no ice cores from then. It may well be that if we had such detailed information for earlier times, it could show dramatic environmental changes at this frequency. But possibly not as ice ages are intrinsically fickle and the ice likes to come and go at the slightest suggestion of a temperature change.

It was during the Quaternary that the genus Homo (essentially human-like animals) developed. During the 1.8 million years there arose at least eight different members of this genus, including ourselves. Calculating the number of types of homo has become a kind of sport amongst palaeontologists, with almost every fossil uncovered requiring the books to be

> *rewritten.*
>
> *We of course are probably the only surviving members of our genus. This plethora of homo (not to mention other) species over such a short time span suggests very much that fluctuating environments favour rapid evolution. At present extinction seems to be outstripping the Earth's ability to replace species. This is usually blamed on humans, but, whilst humans may be the cause of some extinction, it may simply be that we are accelerating a process that would have happened anyway.*
>
> *The driving force behind the alternating glacial/interglacial periods seems to be a link between there being a land-locked Arctic ocean and aspects of the Earth's orbit around the Sun, called Milankovich cycles, which determine which part of the Earth is more or less warmed by the Sun.*

It is not fully understood what feedback mechanisms are working to stabilise the Earth at present but one feature stands out. The Gulf Stream brings warm water from the tropics towards North West Europe keeping the climate there milder than it would be otherwise. The Gulf Stream consists of warm, dense water from the Caribbean (dense due to evaporation in the tropics), which is carried northeast, cooling as it goes. When it gets cold enough, just off Iceland, it sinks to the bottom and flows south again, keeping the cycle going – a bit like a giant conveyer belt – and indeed another name for it is the North

Atlantic Conveyer. It is an important way of transferring warmth from the equator to higher latitudes. During glacial periods the Gulf Stream is much further south.

The irony is that global warming melts ice in the North Atlantic, which dilutes the northern bit of the Gulf Stream. This stops the waters from being as dense, so they don't sink and the flow of warm water stops. If this happens it results in North West Europe becoming colder, despite a global rise in temperature. Of course, warmer global temperatures fairly rapidly melt the ice that has been diluting the Gulf Stream. When the ice runs out, which may be very quick in geological terms, the Gulf Stream is reinstated.

Obviously lots of other feedback mechanisms are in progress and this makes it difficult to work out what is going to happen next, although, ironically, climate is easier to predict than the weather. Weather predictions need a super computer to work out what will happen next and famously the weather in Britain becomes a national obsession, as predictions are so often wrong.

The story that comes out of this is an Earth that is incredibly stable despite the attempts by outside forces to destabilise it. Various ice ages have come and gone. At times the Earth has been very warm compared to the present, with carbon dioxide levels much higher than now. However, through it all liquid water has been present, enabling life to survive - and even thrive.

What I have ignored up to now is the influence of the Moon upon the Earth. A little background is necessary.

The Formation of the Moon

The Moon tends to be taken for granted as it is always there, however it does make a dramatic difference to life on Earth. The formation of the Moon is extremely well understood for something that took place 4.5 thousand million years ago. At this time the proto Earth had 95% its present mass and was orbiting the Sun in its present position. A planetary body, about a tenth the mass of the Earth and travelling in the same orbit, impacted with it at a very shallow angle at about 40,000 kilometres per hour. Both the proto-Earth and the proto-Moon were already differentiated, which means that they had nickel/iron cores and lighter rocky material outside. Importantly, their composition shows that they had both formed in the same region of the Solar system.

There were, at least, two impact events. The first impact stripped off the outer layers of the two bodies, mostly from the proto-Moon, because it was so much smaller. The proto-Moon was then attracted back by gravity and hit the planet again. In this collision most of the core of the proto-Moon was absorbed into the Earth's core, leaving essentially rocky crust material, molten from the heat of impact, in orbit around the Earth. The Earth thus ended up with a larger core than before and the orbiting molten material coalesced to form the Moon, which contained much more of the lighter material than before, with only a very small nickel/iron core (if any).

Initially the Moon was closer to the Earth than at present and has gradually moved out to its present position. In fact it is still moving away from the Earth at a few centimetres a year. This is not really relevant because the Moon is affected by things like the distribution of land masses on the Earth and can move slightly closer or further away, depending on circumstances.

What comes out from considering the origin of the Moon is how unbelievable the circumstances are. The compositions of both the proto- Moon and The proto-Earth in the accepted theory have to be very exact for this to work and the existence of two large planetary bodies in the same orbit is incredibly unlikely because they would be expected to annihilate each other before they were properly formed. Most astronomers talk about odds of over a billion to one against the formation of the Moon in this way. The bottom line is that the Earth/Moon system does exist, however unlikely. It is a bit like the lottery, there are odds of fourteen million to one against winning the lottery and yet people do win it almost every week.

The Moon stabilises the Earth so it does not flip on its axis and is restricted to a very narrow range of angles, from 21.8° and 24.4°. The reason for this stabilisation can best be seen by considering the sport of hammer throwing. In this you have a darned great iron ball on the end of a chain, whirl it round and round and then throw it. While you are spinning round, whirling the hammer, you are more or less vertical, but when you release the hammer you tend to fall over. At least I do.

If the Earth did flip on its axis the result would make other mass extinctions look like minor events.

The collision leading to the formation of the Moon gave the Earth a larger core and thinner crust than it would have had otherwise. Plenty of extra heat (the kinetic energy of the impact being converted to thermal energy) would also have been produced. We have already seen that plate tectonics is essential for there to be continuing life on Earth, as without it there can be only limited recycling of the crust. A thin crust enables plates to break up more readily and a larger core generates higher temperatures and encourages convection in the mantle to drive plate movements. Higher initial temperatures also help this process.

Having a large Moon means large tides, not just in the oceans but also in the atmosphere and the mantle. Without these tides the Earth would stagnate, as there would be reduced flow in all of these.

Overall it is difficult to see how a planet without a large Moon could continue to support life, especially any kind of complex life, for any reasonable period of time.

In this chapter I have presented the necessary conditions for life to exist on Earth with especial attention to the presence of a large Moon to stop the planet from falling over; plate tectonics, as a giant recycling mechanism; and water, some of which would have been liquid in even the most extreme environments the Earth has known. The next question is really, "is the Earth unique"? In the Solar System the rocky planets are the closest analogues of Earth and it is these we visit next.

Where Are All The Aliens

Chapter 5

Life And The Rocky Planets.

The inner rocky planets are the obvious place to look for life, if nothing else because we live on one of them, the Earth. Closest in to the Sun is Mercury and intuitively we would not expect life there as, being so close to the sun, it would be too hot. In fact the surface of Mercury is not the hottest planetary surface; this record is held by Venus. It was thought for a long time that Mercury was in a synchronous orbit with respect to the Sun, with a really hot hemisphere facing the Sun and a really cold hemisphere facing away from it. This is, however, not quite true. In the time that it takes Mercury to go round the Sun twice (176 days) it has rotated three times. (This three to two resonance can be seen in other members of the Solar System, particularly with moons of the gas giants, so is not uncommon.)

The surface temperature on Mercury ranges from minus173°C to 427°C and this wide range tells you immediately that the planet does not have a substantial atmosphere. If it did, heat would be distributed, by the atmosphere, from the hot to the cold side. Mercury does have a very tenuous atmosphere comprised of atomic oxygen (single atoms of oxygen rather than molecular O_2), sodium atoms and molecular hydrogen with some helium derived directly from the Sun. The lack of a substantial atmosphere is due to two factors, the first being entirely due to its position. Mercury's closeness to the Sun means that it experiences a stronger solar wind than the other planets, which literally blows off any atmosphere. Also it would have lost considerably more of its atmosphere during the Sun's T-Tauri stage than planets further away from the Sun.

The Solar Wind and T-Tauri Wind

The solar wind is a stream of particles, mainly electrons and protons that flow out from the Sun mostly following the lines of the Sun's magnetic field. The solar wind, like any other kind of wind, can carry other particles along with it. Hence the loss of Mercury's atmosphere. This wind varies with the eleven-year period that the Sun's activity follows – i.e. the Sun spot cycle.

The T-Tauri stage (named after a star called T-Tauri, which is the classic example) occurs in most medium mass stars, early on in their formation at the point where the star is almost completely formed and any planets are beginning to be accreted. During this stage the star sheds a large amount mass

> *over a relatively short period (several million years) until it stabilises. The matter that is shed forms the T-Tauri wind, which is similar to the solar wind but much more powerful. This strips the lightest gas and dust from the inner stellar system. The T-Tauri wind is one reason why there are no gas giants close in to the Sun. (Other stars do have gas giants in their inner system but these are thought to have formed further out and migrated inward.) Mercury clearly would have had little chance to hang on to its lighter gases. One very important effect of the T-Tauri wind is that it would also have blown away most of the organic matter from the Inner Solar System.*

The other main reason for the lack of an atmosphere on Mercury is simply its mass (Mercury has a mass 5.5% that of the Earth). This much smaller mass, combined with high temperatures, means that gases will be lost as they exceed the escape velocity of Mercury. Significantly, water vapour would not be retained.

> ### *Escape Velocity*
> *All gases are in motion as a result of their temperature. Gases with a low molecular weight have a higher speed than those with a higher molecular weight. The force of gravity on a planet is directly related to its mass and affects everything on its surface, including the atmospheric gases. The radius of the*

> *planet is also important, as on a large planet the atmosphere is further away from the centre of gravity. If a molecule of a gas is travelling straight up from a planetary surface it experiences the pull of gravity, which slows it down. Molecules at the top of the atmosphere have to be heading upwards with a certain speed if they are to escape into space and be lost from the atmosphere and this is called the escape velocity. It applies equally to any object, so a rocket has to exceed the escape velocity of the Earth in order to enter space (for Earth this is 11.2km per second). With gases there will be a spread of velocities with some molecules of a particular gas having a higher velocity than others. Because of this a planet will start to lose a particular gas when its average speed is one sixth of its escape velocity. The lighter gases, being faster, are lost most easily.*

The surface of Mercury has been imaged close up only once, by Mariner 10 in 1974, when 18 images were taken, covering 45% of the surface. Mercury's surface looks very much like the Moon's, being heavily cratered and with evidence of ancient lava flows. One very surprising discovery was made in 1991, when radar images, from the Earth, of Mercury's north pole showed a bright target with the signature of water ice. This ice almost certainly arrived via comets and is shaded from the Sun by the wall of a crater, (the inclination of Mercury's pole of rotation is only 0.1 degrees so this shading is possible). However, the low pressure environment on Mercury means that ice would sublime (go straight to gas), rather than melt to give liquid water and of course the water vapour would be readily lost from the atmosphere. So no possibility of life on Mercury. But it is useful to look at this planet as it gives conditions where life cannot exist, even on a planet that has a degree of similarity to Earth.

Venus, the next planet out from Mercury is, on first glance, the sort of planet to make you think "Wow! This planet is so like Earth that it must harbour life". Indeed, until the 1960s most people believed that life was possible there. Venus is about the same mass as Earth (0.82 that of the Earth) and has a substantial atmosphere. Venus is completely covered in clouds and nothing at all was known about what lay beneath them, allowing people's imagination a free reign. The Russian Venera 13 spacecraft managed a successful landing in 1982 sending back, amongst other data, the first colour pictures of the surface. Venera 13 did not last long because the surface of Venus was found to be outrageously inhospitable.

The USA and the USSR Planetary Exploration During the Cold War

The Cold War lasted roughly from 1947 till 1989 and, after the Russian Sputnik was successfully launched on 4th of October 1957, the USSR and the USA indulged themselves in competing to be the first country to achieve various goals in space. There was naturally a hidden agenda; rocket science can also be used to fire missiles at other countries. (I do hate the phrase "It's not rocket science" implying that rocket science is really difficult. Rocket science is mostly the application of brute force - the bigger the rocket the more you can send up. What is difficult is stopping people firing the rockets at each other.)

Initially the USSR had the lead, landing a space vehicle on the Moon. Ironically, the main reason why the USSR had the initiative was because they were behind the USA in the development of nuclear weapons. Both the USSR and the USA used the expertise of German rocket scientists but the USA, being ahead in the development of nuclear weapons, had smaller and lighter missile warheads, so the USSR was forced to develop much more powerful rockets than America. The USA could mobilise far more resources than the USSR, as their country had suffered very little in the Second World War compared to the devastation inflicted on the Soviets. The Americans very effectively gained the support of their people by such gimmicks as parading pictures of the Moon complete with hammer and sickle (the hammer and sickle being a symbol of the USSR).

With respect to the Moon the USA trumped the USSR with their manned landings using Apollo 11 and 12 in 1969. Although both countries sent spacecraft to Mars and Venus, the USA concentrated more on Mars and the USSR more on Venus. I believe that one of the underlying reasons for this difference originates in literature. Science fiction in the USA, at this time, tended to set their space operas on Mars, whereas Russian science fiction based theirs on Venus. Both however had a very clear idea as to how women on these planets would dress, mainly in leather bikinis. I always felt that they should at least have a warm cardigan and matching leg warmers - far more practical, especially at the temperatures on Mars.

At the end of the day the Cold War did not turn hot, so I am still here to write about it, and much scientifically useful material came out of the space race. With respect to Venus, the Russian Venera landings have given us a good idea of conditions on the surface of Venus and the USAs Magellan craft in 1989 sent back a radar map. At the moment (2007) the European Space Agency (ESA) has the spacecraft Venus Express orbiting Venus, making a much more detailed map of the surface.

As Venus has a very dense atmosphere the pressure at the surface is 92 times that of the Earth. When thinking about this pressure with respect to life most writers seem to think this poses a problem. However, this pressure is equivalent to just a bit less than that at 1km depth in the Earth's oceans, which can reach a depth of almost 11km. At these depths the pressure is over a thousand times that on the Earth's surface, but even here you find a lot of life, (though not as plentiful as at shallower levels).

Pressure then is not that great a problem for life; unfortunately the same cannot be said for the temperature. On the surface of Venus this reaches 460°C, which, we are often reminded, is hot enough to melt lead. The highest temperature at which any organism has been observed to reproduce is 113°C. These organisms are called thermophiles, meaning they thrive at high temperatures. (Thermophiles are part of a motley bag of organisms called extremophiles, which as the name suggests can survive, if not thrive, in extreme environments.) Biological research indicates that 150°C is the limit for life, as above this temperature genetic material simply disintegrates. So the temperature is too high for life to exist on the surface of Venus. Another factor that prevents life existing at the surface is the lack of water, there simply isn't any.

Extremophiles

Extremophiles are organisms that not only live in extreme conditions but also need such conditions to thrive and reproduce. Extreme conditions are of course relative. We would be crushed to death if we descended to the bottom of the Ocean without protection, and therefore consider this environment extreme, whereas most organisms that normally live there would die if brought up to the surface and would consider the lack of any decent pressure in surface waters to be inimical to their kind of life. Most extremophiles are simple organisms, such as bacteria, but there are many other organisms (fish, fungi, insects etc) that require extreme conditions. Some extremophiles require more extremes than one, for example high temperatures as well as acid conditions. It is impossible to list all the extremophiles we know about, as almost everywhere you look on Earth there are extreme conditions occupied by organisms, so the following are just examples. Thermophiles like it hot and are found in hot springs, hydrothermal vents etc. Acidophiles like acid conditions (down to a ph of 0, as in concentrated sulphuric acid,) once again in hot springs and hydrothermal vents. Alkaliphiles like highly alkaline conditions, such as found in soda lakes. Psychrophiles like cold conditions down to about minus 10°C such as Antarctic ice. Xerophiles like dry conditions such as hot or cold deserts. Halophiles like salty conditions such as salt pans. Barophiles

> like high pressures such as inside rocks or at the bottom of oceans. Many more extreme environments exist. I was reading the other day about a bacterium that lived in oil deposits and I suppose you would call this an oleaphile.
>
> Extremeophiles are good to bear in mind when you are looking at extraterrestrial environments, as otherwise it is far too easy to dismiss the possibility of life just because from our viewpoint the conditions are extreme.

One of the biggest surprises from the Magellan probe was the discovery that the surface of Venus is only five hundred million years old, which may seem old from a human viewpoint but geologically speaking it is very young. You must bear in mind that I am not talking about some bits of the surface being one age and other bits being another. All the data strongly suggest that the whole surface of Venus was literally created five hundred million years ago. Not created in the biblical sense of coming from nothing, but created in a planet wide process that destroyed the preceding surface and replaced it completely with molten material from below, which then set to form solid rock.

> ### *Dating Planetary Surfaces*
> *It may look an impossible task to give a date to a planet's surface, especially if no spacecraft has ever landed on it,*

however astronomers are very sneaky people and have come up with a number of ways to accomplish this. Material that is native to the Earth can be dated in lots of ways and amazingly they mostly all give dates that agree with one another. Radioactive isotopes with a suitable half-life (the half-life is simply the time it takes for an isotope's radioactive emission to drop by half) provide a very reliable dating system, especially when a number of isotopes from the same sample are utilised. This gives the date when the material was last molten, as melting and recrystallisation partitions elements (in this case radioactive elements) in a predictable way that has been verified by laboratory experiments. In a sense the radioactive clock is reset when fresh rock/minerals are formed from the melt.

For very old rocks zirconium silicate crystals are used. When a crystal of zirconium silicate (zircon) first forms it is clear, but the Earth is constantly bombarded by cosmic rays and these leave small track lines in the crystal. The flux of cosmic rays (the amount hitting the Earth) is, as near as damn it, constant, so the number of tracks in a crystal is a direct indicator of age. Of course other effects are present, but with enough samples and cross correlation with other evidence, zircon crystals are surprisingly good radiogenic clocks.

For relatively recent age determination (about the last two hundred million years) magnetic stripes on the ocean floors are good timekeepers. Knowing the rate of spreading of the ocean floor (which can be determined with a high degree of accuracy) the age of a piece of ocean floor can be calculated. Two hundred million years is the limit though, as ocean floor returns to the Earth's mantle due to plate tectonics.

Much material from space arrives constantly on the Earth's surface in the form of meteorites. Some of this material is from the asteroid belt (you can tell by comparing it with the observed properties of asteroids), but also some has originated from the planets or the Moon, having been blasted off following impact by an asteroid or a comet. Most of this material comes from the Moon or Mars, as the thick atmosphere on Venus makes it extremely unlikely that an impact event would give enough energy for any rock to escape into space. The other bodies in the Solar System are so far away it is unlikely that any material from there will reach the Earth. This asteroidal material represents stuff that has remained largely unchanged since the Solar System was forming and this is one way that we get an age for the Solar System.

The Moon, of course, is the only extra-terrestrial body that mankind has visited and material collected there gives us firm dates for different parts of the Moon, including some of the craters. The four inner planets would be subjected to much the same bombardment as the Moon, but erosion has taken its toll. If a planet is resurfaced, as happened on Venus, there will be no craters to start off with. It is fairly complicated but the number

> *of craters on a planet depends on the time that it was last resurfaced. Simply put, the more impacts the older the surface. In the case of Venus there are few craters but enough for statistical analysis to give a reasonably good date for the last resurfacing of the planet.*

Now, on Earth the surface is continually being recycled via erosion and plate tectonics. The latter process allows heat from deep within the Earth to be carried to the surface, where it can be dissipated. This needs to be an ongoing process, as the heat comes not only from residual energy from planetary formation, but also from radioactive elements in rock, which are continually warming things up. All rocky bodies in the Solar System are producing heat this way, but it can only build up in the larger ones.

The usual explanation for the resurfacing event on Venus is that plate tectonics do not operate there, at least not at the moment, so the only way that Venus can lose its internal heat is by conduction through the lithosphere (the solid rocky outer shell). Rock is a very poor conductor and so conduction is a much slower means of losing heat than convection, which is how the Earth loses heat via plate tectonics. Because the heat is lost slowly it builds up underneath the lithosphere, partially melting it, until it becomes lighter than the overlying lithosphere and overturns.

I have some difficulty with this explanation. Because there is no efficient means for the heat to be transferred around the mantle (the molten bit below the lithosphere) pockets of heat would build up in the absence of convection. Without sufficient heat transfer the lithosphere would be overturned locally which is not what we observe.

As well as having a relatively new surface Venus has the slowest rotation period of any other planet, 243 days. Curiously, this rotation is retrograde, i.e. it is in the opposite direction to that of the Earth. As Venus orbits the Sun in 224.7 days its day is longer than its year! Another result of this slow rotation is that Venus has no magnetic field. (The magnetic field on Earth is caused by the spinning of the molten outer core.) An explanation for this slow, retrograde rotation is that Venus probably suffered a major impact early in its history. Astronomers are a bit like archaeologists: when archaeologists make a find they cannot explain then they call it a ritual site; when astronomers find features that they cannot explain they call it the result of a collision. The impact theory explains the low rotation, to a certain extent, but leaves many questions unanswered. To have enough energy to virtually stop the rotation of Venus an impacting body would have had to be huge, a huge impact means a lot of debris thrown up, not all of which would fall back to the surface of Venus and at least some would be left orbiting the planet. Venus is conspicuous in not having a moon or a ring of debris around it.

More aesthetically pleasing is a mechanism that explains both the young surface and the very slow rotation of Venus. All planets rotate about an axis, which is tilted to the plane of their orbit about the Sun; the tilt of this axis of rotation varies. In the absence of any outside stabilising influence, the direction of this tilt becomes chaotic. Chaotic in this sense just means that a small influence, for example a volcanic eruption, can have a large effect that cannot entirely be predicted - much like a spinning top. Earth is stabilised by having a large Moon, which provides a continual drag on the Earth, slowing down any change in tilt. The axial tilt of Mars, with only two very tiny moons, has been shown to vary wildly. One of the almost inevitable consequences of this chaotic variation is that at some point the planet will flip over completely, again just like some spinning tops.

If Venus has flipped over, as suggested by the evidence, the atmosphere, the liquid mantle and the solid lithosphere would adjust at different rates, generating friction between them. The result of this friction would be to slow down the rotation of the planet as rotational energy is converted into thermal energy. This released thermal energy would heat up both the lithosphere and the mantle on a planet wide basis, causing the observed resurfacing of the planet all at the same time.

There is another curiosity to do with the rotation of Venus. Venus rotates three times on its axis (729.27 days) in almost the same as the Earth takes to rotate around the Sun twice (728.50 days), so at closest approach the same face of Venus is always opposite the Earth, with very small variation. This sort of 3 : 2 resonance is common among the moons of planets and usually indicates tidal influences. It seems unlikely that tidal effects, at the distance of the Earth from Venus, would have much affect on a solid planet, but it is also unlikely to be a coincidence, simply because it is too exact and coincidences don't happen without a reason. However, if the scenario I have suggested for the resurfacing of Venus is correct then the enormous energy released from the frictional slowing down of Venus could easily have melted the whole planet. A molten Venus could distort much more easily towards the Earth and the tidal bulge formed would have acted as a further brake on the rotation of Venus after it flipped over. Whatever the precise mechanism, the resurfacing would have erased any life forms on the surface of Venus that may have developed before the planet heated up.

The atmosphere of Venus has about 92 times the mass of Earth's atmosphere. The composition is about 97% carbon dioxide with most of the rest being nitrogen with clouds that are mainly composed of sulphuric acid and water (75% sulphuric acid, 25% water). The amount of carbon dioxide would result in a runaway greenhouse effect, which accounts for most of the high temperature on the surface (the rest being mainly accounted for by the escape of internal heat).

> ### *The Greenhouse Effect*
>
> *All bodies that are not at absolute zero radiate electromagnetic energy. The frequency of this radiated energy depends upon the temperature of the body, the hotter the body the shorter the wavelength. (The electromagnetic spectrum starts at the shortest wavelengths with gamma rays, x-rays, ultraviolet, visible light, infrared, microwaves and finally radio waves, which have the longest wavelengths). The Sun, with a surface temperature of 5800K, radiates most of its energy in the visible part of the spectrum. When a planet is heated by the Sun some of the energy is used to heat up the surface, so it reradiates at a longer wavelength, in the infrared part of the spectrum. Various gases, among them carbon dioxide, water vapour and methane absorb energy in the infrared part of the spectrum and then reradiate it. Incoming radiation from the Sun comes from one direction, reradiated radiation is emitted in all directions, so much is redirected back down through the atmosphere towards the surface and the rest of the atmosphere, warming it further. Carbon dioxide in the Earth's atmosphere is only about 350 parts per million by volume, compared to about 210 ppm before the industrial age. Even at these very low levels carbon dioxide is contributing to the warming of the Earth.*

> *Put into this context it is easy to see why Venus is so hot. The vast amount of carbon dioxide in the atmosphere acts like a thick blanket, or a very high TOG duvet, keeping the heat in. Most of the difference in atmospheric carbon dioxide levels between the two planets is the result of carbon dioxide being buried in rocks and organic deposits on the Earth but not on Venus. However, at the moment, we are trying to change this situation by burning these deposits of coal and petroleum and so returning the carbon dioxide to the atmosphere.*

Although the surface and lower atmosphere of Venus is hot it does cool down in the upper reaches, so at about 50 kilometres above the surface the temperature is about 70°C. This is the level where the clouds exist and these clouds are a possible habitat for acidophiles.

In these clouds hydrogen sulphide and sulphur dioxide have been detected. These two compounds react with each other, so should not be present together. The obvious (though not necessarily correct) conclusion is that life forms are producing these gases. Furthermore, there is virtually no carbon monoxide in these clouds where there should be lots from the photolytic dissociation of carbon dioxide so something is removing it. Carbonyl sulphide is also present with no indication of a non-biological source. If there are acidophiles in the atmosphere of Venus they could be utilising ultra violet radiation as an energy source. Ultraviolet pictures of Venus show dark patches where the ultraviolet radiation has been depleted.

Such hypothesised organisms have to come from somewhere else, as the concentrations of relevant chemicals in the clouds, while enough to maintain life, are probably not enough for them to originate there. Early in its history the Sun was considerably cooler than at present. It is quite likely that primitive life could have developed on the surface before it became uninhabitable and then survived by migrating to the cloud levels. A number of organisms have been shown to inhabit clouds and the upper atmosphere of the Earth. There are plans to sample the clouds of Venus for life and there is a good chance they might succeed, however micro organisms do not count as intelligent communicating life.

Skipping the Earth, dealt with in the previous chapter, we come to Mars, which is now the favoured site to search for life. Mars is the most visited extra-terrestrial body other than the Moon, so we should have an awful lot of information about it. Most of the early missions however were a complete disaster, with some probes getting lost on the way, some failing to achieve orbit, some crash landing and others not functioning as planned. Europe, Russia and America have all had a go and all had failures – but in recent years there have at last been some successes.

The surface of Mars is very dry and cold, with average temperatures of around minus 50°C. The atmosphere is made up largely of carbon dioxide and nitrogen and is very thin, giving a surface pressure only six thousandths of that on Earth. Pictures from the Landers have shown us a bleak, barren scenario, covered in reddish rocks and dust. We know that planet wide dust storms are a regular event, happening about twice in a Martian year. Substantial cratering implies that some of the surface is very old, though the lack of craters in other parts suggests some renewal of the surface, with some parts as young as a million years old, very young in geological terms.

The most noticeable feature on the surface is Olympus Mons, a huge, probably extinct, shield volcano. This is the largest volcano in the Solar System and stands 24 kilometres high, three times taller than Everest. The very existence of Olympus Mons tells us a lot about the internal processes of the planet. The process of plate tectonics on the Earth continually moves the crustal plates around the surface. Mantle plumes, carrying heat and magma from the depths, cause volcanoes on the surface, but as the plates move you get a chain of volcanoes, as in the Hawaiian Islands. With no plate tectonics on Mars Olympus Mons just grew and grew.

Mars, you will remember, is much smaller than the Earth, having just over a half the diameter and only one tenth the mass of our home planet. Being so much smaller Mars would have lost its heat of formation much more efficiently and would have fewer, heat producing, radioactive elements. There would therefore be less volcanic activity and so less outgassing to replenish the atmosphere. As the planet cooled the lithosphere would have thickened, making it more difficult for mantle plumes to penetrate from below. Existing volcanoes would therefore be the only conduits for heat to reach the surface – and these would grow. A thicker lithosphere can also support more mass. On Earth, Olympus Mons would sink right through the crust.

Unlike Earth, Mars does not have a global magnetic field and hasn't had one for the past three and a half thousand million years. Earth's magnetic field channels away cosmic rays and the solar wind. Without this protection Mars is continually bombarded, effectively sterilising the planet's surface. In addition, with no oxygen to speak of there is no ozone layer and so no protection from harmful ultraviolet radiation. Future manned missions had better remember the sun block.

For the sake of completeness I should mention Phobos and Deimos, the moons of Mars. Unlike our moon, these are captured asteroids and far too small to have any noticeable effect on the planet.

It's safe to say that the present surface of Mars does not look like a good bet for life of even the most primitive kind, though this doesn't stop people looking for it. In the past, however, Mars would have had a denser atmosphere, largely carbon dioxide, which, even with a weaker sun, would have kept the planet warm enough to support life. The Viking Landers sent by NASA in 1975 were specifically designed to search for life, but to be honest rather messed up and their results were ambiguous. The surface of Mars is exposed to ultra-violet and cosmic rays, which lead to chemistry not anticipated by the planetologists involved. However, with NASA's two rovers and ESA's (European Space Agency) Mars Express orbiter along with other craft, substantial new information is available. Obviously water would need to be present for life to evolve so it comes down to the question of whether or not water existed on Mars long enough to sustain life. The possibility of water on Mars has been one of the main driving forces behind the exploration of the planet and there is now abundant evidence that not only was there once water on Mars, but that it is still there, buried under the surface.

Water on Mars

The surface of Mars is totally dry, if nothing else because the atmosphere is too thin and the temperature is too low. It came as a surprise when, in 2002, NASA's Mars Global Surveyor sent back pictures showing features that seemed to indicate that water had not only flowed in the past but may have flowed very recently. Features on the surface include channels identical to those caused by flowing water on the Earth. Various scenarios were suggested that did not involve water, such as dust flows and carbon dioxide ice, but no mechanism that didn't involve water was particularly convincing. Then in 2004 one of the NASA rovers found rock formations that were clearly sedimentary and minerals that are known to only form under water. A radar scanner on the Mars Express detected water ice in large quantities at the North pole of Mars as well as some away from the poles. It is likely that there is liquid water underneath the ice, as it will be warmed from below by geothermal sources At the end of 2006 a feature looking like the results of flowing water was discovered on a photograph of Mars, which had not been on previous photographs, so watch this space

. Of much more interest is the detection of methane on Mars, as methane would not last long in the planet's atmosphere, so must be being replaced. Life is the obvious candidate as the methane source, although some astronomers consider it to be a sign of volcanic activity taking place at present. Even if the methane is volcanic in origin it indicates that there are warm areas on Mars that could harbour life.

So, Mars does offer the possibility of life living in underground pockets of water, though once again it can only be very simple life, but still interesting. It should be noted that any life on Mars is likely to be halophilic (salt loving), as any buried water will be extremely salty due to the overlying ice rejecting salts from itself as it formed.

Although the asteroids are not planets in themselves they do interact with them. Once upon a time it was thought that there had been another planet in an orbit between Mars and Jupiter, which had suffered a cataclysmic disaster and broken up to form the asteroid belt. But it is now recognised that no planet could have formed here because of the influence of nearby Jupiter, destabilising any largish body. No-one has suggested that life exists on the asteroids, but they may have played a part in delivering organic materials to the inner planets. As the asteroids formed further out in the Solar System they retained more of their organic materials during the Sun's T-Tauri phase than did the rocky, inner planets so they may have had an important role in the story of life by helping it to arise on the other planets.

So with the rocky planets we cannot conclusively reject the possibility of life in the atmosphere of Venus or buried under the surface of Mars. It is important to realise that such life will, of necessity, be primitive and not the intelligent life we are looking for. Intelligent life requires time and a dynamic environment to evolve and develop; such an environment is not available where life can only cling on. Until recently I would have stopped here in the search for life within the Solar System, but I have had my ideas turned on their head by the exploration of the gas giants, which I consider next.

Where Are All The Aliens

Chapter 6

The Gas Giants

Unlikely as it seems, considering how far away they are and how unlike the rocky planets, we have a reasonable idea of the composition of the gas giants. The first assumption must be that these planets have a similar composition to the original nebula from which the Solar System condensed. Initially this looks unlikely to be true, as Jupiter and Saturn are not far enough away from the Sun to hold on to their original atmospheres during the T-Tauri wind, early in the formation of the Solar System. The effect of this strong stellar wind would have been to strip the lighter gases from the atmospheres of the planets out to quite a considerable distance from the sun.

In fact the consensus is that Jupiter and Saturn formed further out in the Solar System and quite early on migrated inwards, whereas Uranus and Neptune formed closer and were thrown outwards. There is a difference between the four gas giants in their makeup (see below), which also fits with them having migrated. As you will see later migration of planetary orbits around stars seems to be a common phenomena. Spectroscopy tells us directly what the outer atmosphere of the gas giants consists of and their mass and density can be easily calculated from the orbits of their moons and any passing spacecraft. Laboratory work on the Earth can duplicate quite a lot of the conditions likely to be found in the interiors of these planets.

> ### *The Composition of the Gas Giants*
> *This might seem a bit boring but sometimes you have to resort to numbers to get the information over. You needn't bother trying to remember any of this (unless you want to), so I have tried to keep it short. Jupiter has a mass 318 times that of the Earth and a density just less than a quarter that of the Earth. Saturn has a mass 95.2 that of the Earth and a density about an eighth that of the Earth. Uranus has a mass 14.4 times that of the Earth and a density about a quarter that of the Earth. Finally Neptune has a mass 17.1 that of the Earth and a ideensity just over a quarter that of the Earth.*
>
> *The compositions are somewhat variable, with Jupiter and Saturn being similar to each other, as are Uranus and Neptune. Jupiter and Saturn have rocky/icy cores, which are probably differentiated (more dense rock in the middle and less dense rock/ice around this). These cores are about five Earth masses*

for Jupiter and twelve Earth masses for Saturn. Above the core both planets have a mixture of helium and metallic hydrogen (metallic hydrogen only exists under high pressures, greater than two thousand times the atmospheric pressure at the Earth's surface). This layer is five times as massive on Jupiter as on Saturn and largely accounts for the major differences in density between the two planets. The topmost layer of these planets consists of a mixture of gaseous hydrogen and helium with, interestingly, some rocky material. Only extremely tiny traces of water have been confirmed in the atmospheres of Jupiter and Saturn. There may be some more water below this, but it will not be liquid under the extreme pressure regimes. There are also traces of some simple organic compounds.

Uranus and Neptune are rather simpler. They both have rocky cores of about one Earth mass, then an icy layer (largely water, ammonia and methane) topped with a hydrogen and helium atmosphere, with a few traces of other compounds. Pressures for the gas giants are extremely high even when approaching the edge of the atmospheres and temperatures are higher than you might expect considering their distance from the sun. All the gas giants have significant magnetic fields, although those of Neptune and Uranus are tilted considerably away from the axis of rotation of their respective planets; this has been explained by them having only a shallow conducting fluid layer in their makeup.

The extreme scarcity of organic materials and liquid water on Jupiter and Saturn means there is absolutely no hope for any kind of complex life developing. Uranus and Neptune may be more endowed with the appropriate materials, but they are so far away from the sun that the temperature is just too low for the chemical reactions to take place. So, no hope of life on the gas giants, though I'm sure that won't stop astrobiologists looking.

What the gas giants do have are moons, lots of them, and some of a substantial size. At the beginning it is worth recognising that the existence of any habitat for life on a moon of a gas giant is highly speculative. It could be said that, to justify the search for life in the Solar System, scientists sometimes have a tendency to scrape the bottom of the barrel rather a lot, after all careers are at stake. I don't hold with this entirely, as the discovery of only one, fully attested, extraterrestrial organism would revolutionise the whole way we look at the Universe.

Most of the moons of the gas giants have heavily cratered surfaces, implying they are very ancient and have no atmosphere. These aren't worth mentioning in terms of the search for life. Others however are intriguing in the extreme. Now, we know that the number of craters on the surface of a planet or moon has often been used to date the surface, thus giving information about planetary history. At this stage it is worth considering the limitations of this method for the moons of the outer planets.

> ### *Crater Dating for the Moons of the Outer Planets*
>
> *The number of craters on the Moon and where they are found has been used to calibrate ages for the rocky planets. We have actual samples of Moon rock, which we can accurately date. With Mars we can extrapolate the data from the Moon to give ages of particular parts of the surface. Dating of the Martian surface is helped by analysis of meteorites, originating from Mars that have been found on Earth. However, there is a fairly large measure of uncertainty due to the scarcity of Martian meteorites on the Earth and the fact that Mars is much closer to the asteroid belt, so you would expect more impacts.*
>
> *For the moons of the gas giants we have no material to analyse on Earth. It is probably true that the rate of crater formation was higher in the past than at present simply because, for all intents and purpose, bodies that can cause cratering, originated within the Solar System and so got used up over time. The general rule remains; a heavily cratered surface is older than a less cratered surface. The position of the gas giants puts them well outside the asteroid belt so large impacting asteroids are unlikely to have been the main source of cratering as they were in the inner planets. One obvious source for potentially impacting bodies is the Kuiper Belt, out beyond Neptune's orbit.*

Comets have actually been seen to hit the gas giants, rather spectacularly so when Shoemaker-Levy 9 crashed into Jupiter as the world watched in 1994. Another source (probably the major source) of crater forming bodies is the material that orbits the gas giants. Many people tend to think of the Solar System as having neatly orbiting bodies, bar the occasional wayward comet or asteroid. However, the Solar System does not work this way and is full of all sorts of bodies playing to their own tune. The gas giants gravitationally attract stray bodies as a hobby, particularly massive Jupiter. Another factor to take into account is that almost certainly the gas giants formed further out in the Solar System and migrated inwards to the orbits that they occupy now. As they moved inwards they scooped up all sorts of space junk, much of which is still in orbit about these planets. With such a lot of bodies in orbit collisions are inevitable. The evidence points to impacts occurring at frequent intervals even today.

The downside of this is that dating the surface of any of these moons is impossible beyond saying one surface is more recent than another, and then only if they are orbiting the same gas giant. In fact you would expect heavy bombardment more recently than for the inner planets, with the subsequent heating and overturning leading to younger surfaces. The upside is that when we come across a lightly cratered moon orbiting one of the gas giants we can, without reservation, declare it to be very

> *young indeed. Heavily cratered surfaces simply cannot be dated with our present knowledge but we can, at least, make a judgement on their relative ages.*

Jupiter has in excess of sixty moons, the four largest being Ganymede, Callisto, Europa and Io. Among these moons Europa has often been considered to be the best place to look for life. Europa is just a little bit smaller than the Moon and has a density similar to the rocky planets. It also has as substantial magnetic field, this implying an iron rich core and a rocky mantle. The surface is mostly water ice and has very little atmosphere. The intriguing thing about Europa is its surface, which is very complicated. Firstly the surface reflects light very efficiently, giving an albedo of 0.7.

> ### *Albedos*
> *Albedos can be a useful way of looking at the surface of planetary bodies as they give some idea of their composition. The albedo of a body is simply the proportion of light that it reflects (usually calculated for a particular wavelength or series of wavelengths). A body with an albedo of 0 does not reflect any light at all, whereas a body with an albedo of 1 reflects all light that falls on it. Clearly an albedo of 1 is not possible; it would have to be an object with a mirror finish and even that would absorb at least a little light. An albedo of 0 is also impossible,*

> *as any surface will scatter incoming light at least a little. The albedo for the Earth is 0.3 and for the Moon, 0.07. The albedo for Europa at 0.7 is thus very high for a planetary surface.*

This albedo for Europa, along with other observations, shows that the planetary surface is mostly covered with water ice, possibly less than 1km thick. A young surface is likely, as you would expect that the ice would become dirty through collecting some of the dust that orbits Jupiter. There are few craters on Europa, which also indicates a young surface. Up close the surface comprises of ridges, sheets domes and fissures, but don't get the idea that we are talking about a rugged terrain. The maximum height of features on Europa is about 200 metres, and these heights are unusual for this moon. Topographically it's a bit like the Norfolk Broads, though somewhat colder.

Some scientists analysing the images returned from Europa say that there are many features that resemble plate tectonics. A nice thought but I get the feeling that whoever put out this comparison has never seen Arctic ice floes, which are a much better analogy, although ice at the temperatures experienced on Europa (minus140°C at the equator and minus190°C at the poles) does act more like rock than we are used to on Earth. This low temperature and the virtual vacuum on Europa's surface preclude photosynthetic organisms. The low temperature means that any chemical reaction, including those essential for life, will proceed extremely slowly. The lack of atmosphere would expose any organism to lethal levels of cosmic rays and ultraviolet radiation. The ice cover on Europa is fairly homogenous, more or less the same wherever you look. This leads to the simple conclusion that the surface ice has very much the same density wherever you look. Plate tectonics needs subduction zones, i.e. places where denser material pushes down under less dense material, to drive the process. Otherwise all the plates will just grind to a halt. There are no signs whatsoever of subduction zones on Europa. What we do have is areas where underlying icy sludge has been forced to the surface in a geological process called cryovulcanism, unknown on the Earth.

Cryovolcanoes

As the name implies, Cryovolcanoes are the icy equivalent of normal (rocky) volcanoes. Normal volcanoes are driven by partially melted rock rising through the mantle. When this rock gets to a certain depth it decompresses, lowering its melting point and causing it to become fluid. This then emerges on the surface as lava. In the case of a cryovolacano, a semi-frozen slush rises to the surface decompresses and erupts as an icy sludge. The behaviour of cryovolcanoes is remarkably similar to that of normal volcanoes. There is evidence of cryovolcanoes on many of the moons of the gas giants.

The existence of cryovolcanoes strongly indicates that beneath the surface ice conditions are warm enough for liquid water to exist. Europa does have crustal plates, (probably better to call them floes), that move with respect to each other. They can move either away from each other or parallel, but at most for a few kilometres. (Compare this with long lasting plates on the Earth's surface, where plates exist for millions of years and move thousands of kilometres). The surface does show areas, appropriately called chaos regions, where surface features are interacting, with older features being erased by younger features, even partially overriding each other, and so on, so the surface is continually changing. This dynamic system can only be accounted for by there being a liquid or semi liquid layer beneath the surface that is convecting. Other observations support this. At the distance of Jupiter from the Sun, and particularly on a small body, this liquid layer cannot be maintained by warming from outside and so must be internally created. Tidal forces are the most likely option.

Tidal Heating

Tidal heating quite often happens with satellites of the gas giants, principally because they tend to have a large number of moons. In a simple binary system such as the Earth/Moon system the satellite moon remains in a synchronous, nearly

> *circular, orbit which means that the forces on the moon are mostly constant and do not vary a lot with time. In a system where other satellites exist, the forces on a particular moon are not only due to the planet they are orbiting but also arise from interaction with the other moons, often involving resonances, such as one moon orbiting the planet once whilst another orbits it twice . In its most straightforward guise a moon will take up an elliptical orbit. At the extremes of this orbit the gravity of the planet will try to pull the moon first one way then another. A moon with a rocky core will have its core constantly being pulled around and the friction involved will generate heat.*

Tidal forces on Europa provide the energy to warm the rocky mantle, which explains the liquid/semi-liquid layer that must be present. This gives the possibility of hydrothermal vents similar to those found on the Earth and bear in mind that these are places that life may have begun on Earth. If hydrothermal vents do exist on Europa we can assume that the chemical composition of the emitted fluids is not hugely different from that found on the Earth, consisting of soluble minerals in superheated water. Chemicals in any Europan oceans would be more concentrated than on Earth due to the ice cover, which would reject salts. A concentrated solution may mitigate against life but, to be honest, we do not know enough to be able to say one way or another. The extent of any ocean on Europa is not known but the planet wide existence of chaos areas would suggest that any ocean might also be planet wide, because of the need for a convecting liquid layer beneath the surface. This would be necessary for life to persist as otherwise chemicals may be used up locally and life would die out.

If life does exist in the proposed Europan oceans then it may even be quite complex, but with very low temperatures and no atmosphere at the surface there is nowhere for life to go. Low temperatures mean that any chemical reactions will proceed much, much more slowly (as a rule of thumb chemical reactions double in speed with every ten degrees Celsius rise in temperature) so any organism that evolved in a warm spot underneath the possible oceans would have a huge problem reaching the surface. Apart from the low temperature the fact that the surface of Europa is virtually a vacuum is an impossible hurdle for any complex life form without protection.

Much the same argument put forward with respect to Europa applies to the other icy moons of Jupiter and the other gas giants that experience tidal heating. Moons that are not subjected to tidal heating simply have no energy source for life and are effectively dead.

Although not a candidate for life (it has no significant atmosphere or water) Io stands out from the other moons of Jupiter in that it is the most volcanically active moon in the

Solar System. Io orbits very closely around Jupiter and so experiences intense tidal forces, which heat its interior so much that the heat flow to the surface of Io is thirty times that of the Earth, with the concomitant greater concentration of volcanoes. More than five hundred volcanoes have been recorded on this small moon with over ten erupting at any one time. Sulphur is a notable constituent of Io's eruptions, turning the planet largely yellow and orange, but they primarily emit molten rock.

Saturn has over thirty moons, nearly all of which orbit the planet in the same plane as the spectacular rings. They are located inside, outside and among the rings and the largest, Titan, is larger than Mercury and has an outside orbit.

Of recent interest is Enceladus, another icy moon. This moon has been found to be emitting geysers of ice particles and water vapour, along with various organic compounds, at a high rate (incidentally making it the most cryovolcanic body in the Solar System). It has been estimated that the ice cover on Enceladus is probably less than two hundred metres thick. In terms of the search for life this would make it a prime target, as it would be the easiest icy moon to obtain samples from its interior.

Titan is the largest moon of Saturn and also the only moon in the Solar System to have a decent atmosphere. This is four times the density of Earth's atmosphere and has shrouded the moon in orange clouds, preventing any surface observations. In January 2005 the Cassini spacecraft, sent to orbit Saturn, arrived at Titan and sent the Huygens probe down onto the surface. This probe sent back data for a few hours and Cassini continues to make observations each time it flies by. Thanks to this we have a lot of information about Titan, which is still coming in.

The atmosphere is unlike that of any other body in the Solar System, being composed of about 95% nitrogen with about 5% methane. There are a lot of other minor constituents of the atmosphere, notably a large variety of organic molecules. The methane is something of a mystery as at the levels present all of it would be lost in ten or twenty million years due to photolytic dissociation, so something must be replacing it. No substantial bodies of liquid methane have been detected on the surface so the best candidates seem to be cryovolcanoes or seepage from underground water/ammonia lakes.

The large variety of organic molecules in the atmosphere is also a bit of a puzzle, as chemical reactions would proceed very slowly indeed at the temperature found in Titan's atmosphere. The source of these molecules is probably photochemical reactions starting with methane. (Photochemical reactions occur when ultra-violet light catalyses reactions of molecules, causing them to break apart and reform in different ways. These reactions can occur at low temperatures where straightforward chemical reactions may not occur.) The products of the methane reactions can themselves react further leading to fairly complex organic molecules.

The surface of Titan is deceptively Earth-like having hills, plains and dunes, with boulders scattered around. There are drainage channels, clouds and rain. However Titan has a surface temperature of about minus 178°C. The hills, plains and boulders of Titan consist of water ice, which at these temperatures would be better described as a rock. The dunes on Titan are extensive. It is not yet clear what they are made of, but the two favourite candidates are water ice or organic matter. Water ice seems most likely as Titan's surface has plenty available acting as rock and it is easy to see how this could be eroded by flowing liquid. The drainage channels would seem to have been cut by periodic flows of liquid methane, although, as said earlier, the lakes or oceans that were predicted, prior to the Cassini/Huygens mission, have not been incontrovertibly located. There are, however, indications of at least some small bodies of liquid on the surface. The Huygens probe picked up a whiff of methane on landing and may well have landed in the equivalent of a bog, with the methane liquid under a weak surface. The clouds and rain are primarily of methane. Titan may well have seasonal rainstorms (perhaps the correct word is methanestorms) however, which could temporarily create lakes of methane.

The fact that Titan looks, on the surface, so like Earth gives a hint that similar processes, under different conditions, give similar outcomes, although making this a general rule would be going too far. It has been suggested that Titan is effectively a frozen time capsule of the conditions that existed on the early Earth and that it would come alive when the Sun becomes a red giant star, something that will happen in about five thousand million years. This is a bit fanciful as, over that time scale, Titan would have run out of activity. By this I mean that the energy source that drives the geology would be drastically curtailed, as over time Titan would become locked into a synchronous orbit around Saturn, much like the Moon around the Earth, and the tidal forces would almost disappear. This will not happen with the moons around Jupiter, as there are five large moons comparable in size with each other. These all influence each other's orbits and this destabilisation will prevent any of them achieving a fully synchronous orbit. Titan, however, completely dominates the other moons of Saturn, so will lose this energy source. Another important factor is outgassing. The source of the methane will run out long before the Sun turns into a red giant.

Titan does however have an atmosphere that resembles that of the early Earth and feasibly life could arise, if the planet warmed up by some means. We could speed things up by dropping a small black hole onto Saturn, which would work its way to the centre and initiate nuclear reactions (okay I admit this idea is not that original as something very similar happened to Jupiter in '*2001 A Space Odyssey*'). There is a minor problem as we do not know how to make a black hole, nor have we yet observed small black holes, but the theory is valid, as far as we know. The other problem is that conservation organisations, such as the Friends of Saturn, would probably oppose the plan and they would probably be joined by astrologers as I'm sure Saturn's cosmic influence would be disrupted by turning it into a mini sun.

As has been said before the outer two gas giants, Uranus and Neptune, are different from Jupiter and Saturn. Their moons are generally less dense than the moons of Jupiter and Saturn, but have a largely similar composition. They are all mostly icy and as well as water ice, they contain ammonia and methane and are likely to have rocky cores. Most of the information about these outer gas giants and their moons comes from the Voyager 2 spacecraft, launched in August 1977, which gave them a flying visit ten years or so later. Otherwise the vast distances involved have prevented much useful data from being gathered. Thus not enough is known about most of these moons to draw any sensible conclusions about subsurface liquid water and hence possible refuges for life. The moons of Uranus are rather titchy, the largest, Titania, having a mean diameter of just 789 kilometres. In general, smaller moons are less likely to be active as they dissipate heat more readily, so we should not expect much of interest from the moons of Uranus. Making such a flat statement is inherently dangerous, of course. Many times astronomers have classed objects in the Solar System as essentially boring, only to be (pleasantly) proved wrong when better observations found totally unexpected and very interesting phenomena.

Neptune's largest moon Triton is very interesting; first of all it is big, being 40% larger than Pluto. It also has a higher density than Titan implying an essentially rocky body. The presence of Triton orbiting Neptune does not fit with most accepted theories of how moons form around planets, simply because of its size it should not be there at all, as there would not have been enough left over material from the formation of Neptune to form a moon that size. The conclusion must be that Neptune captured Triton at some point after they had both fully formed. The capture hypothesis itself causes problems, as a collision is much more likely than a capture, but be that as it may, Triton does orbit Neptune. Cratering on Triton indicates a fairly old, varied surface, which has been shown by spectrographic analysis to consist of icy nitrogen, methane, carbon dioxide, carbon monoxide and water. The lack of activity on the surface implies that it is, essentially, a dead moon.

Out beyond the gas giants lies the Kuiper belt and the Oort cloud. The Kuiper belt contains Pluto, its main moon Charon and an increasing number of other similar bodies that are constantly being discovered, leading to a rather irrelevant argument as to whether they should be considered planets. We don't know very much about these bodies as they are so very far away and most of them have only been discovered in the last few years. All those that have been investigated have very low densities, implying that they are effectively composed of ice (not necessarily water ice).

Much more importantly the Kuiper belt is the source of the short period comets, which carry water ice and organic material into the inner Solar System. There are good arguments that they may have seeded the Earth (and other planets) with the materials to build life. This of course also implies that long period comets from the Oort cloud could also have been involved. A more radical concept called Panspermia suggests that viable organisms could drift through space to seed planets with life and comets may have a role in this. What is known is that material can be ejected from various bodies to land on other planets, but whether viable life forms could be carried this way is doubtful because of the high radiation levels in interplanetary space and the time it takes for rocks to transfer from one planet to another.

Comets may seem to be a possible origin for life as they contain water ice and various organic compounds. As a comet approaches the Sun the ice melts and organic compounds can react with one another. Short period comets may have this happen a few thousand times before they finally break up. However, the total amount of time that conditions prevail that could possibly support life is cosmically very short and the surfaces of planets have had a lot more time for life to arise. The long period comets from the Oort cloud have even less time for life processes to start.

I can't really understand why people want life to originate somewhere else and be transported to the Earth. Even if we do not know the details, the early Earth had plenty of opportunities for life to originate.

So, despite their huge distances from the Sun it is just possible that life may exist on some of the moons of the gas giants. It should be emphasised that such life is likely to be marginal as the available niches are so restricted. As will be shown in the next chapter the Sun has a major influence on where life could exist in the Solar System.

Chapter 7

The Sun And The Solar System

The Sun is a fairly obvious member of the Solar System, containing as it does about 98% of its mass, although if you were out in the Kuiper belt or the Oort cloud it would appear as just a particularly bright star. All the bodies so far considered orbit the Sun and are more or less in the ecliptic plane with the notable exception of long period comets from the Oort cloud.

All known life (which at the moment means life on Earth) relies on the Sun to supply energy so that it can exist. The energy given out by the Sun is prodigious at about 150 million, million KW. At this stage anyone writing about the Sun's output tends to compare it to chemical sources e.g. coal, oil, gas etc. so I won't. Although if you do work it out it is clear that, if the Sun used chemical energy, it would have burnt out millions of years ago, or be a lot younger than all the evidence shows. Actually some creationists make this assumption and use this "fact" and the Bible to declare it to be a "scientific" proof that the Earth was created on October 22nd 4004 BC, probably in the morning, as calculated using biblical dates by the 17th century bishop James Ussher (of course this would be Eden time which, as Eden has been located in the Middle East, would be the middle of the night in the USA and various different times of day around the World)!

The energy of the Sun is derived from nuclear fusion. Although the details are a bit complicated, overall most of the energy comes from fusing four hydrogen nuclei together to form one helium nucleus. The mass of four hydrogen nuclei is just a bit larger than that of one helium nucleus and this extra mass is given off as energy. Sometimes you may come across descriptions of the Sun as 'burning' hydrogen, but this is a misnomer, as the fusion of hydrogen has nothing to do with burning, which is a chemical not a nuclear process. I will try to avoid this bad habit, although it is a difficult habit to break.

The energy output from the Sun spans all wavelengths from gamma rays to radio waves but peaks in the visible part of the spectrum, although at times the Sun's relative output at longer and shorter wavelengths increases dramatically. It is actually surprisingly difficult to measure how the total output of the Sun varies. Ground based estimates are complicated by various factors. The Earth's atmosphere absorbs strongly at frequencies either side of the visible spectrum making accurate measurement impossible, cloud cover and haze reflect and scatter sunlight and the temperature of, the atmosphere over the seasons affects the depth of, and hence absorption of the atmosphere. Space borne measurements have only been possible for the past few decades and correlating data from different satellites using different mechanisms to measure the Sun's output is not straightforward. The short time that satellites have been in operation means that interpretation of anything, other than short-term fluctuations, is not feasible. There is one easily observed phenomenon on the Sun that gives a qualitative measure of the Sun's output and that is the number of sunspots.

Sunspots

Sunspots are dark spots on the Sun representing relatively cooler areas (they still have temperatures of about 4600K). They have a lifetime of a few weeks and follow an eleven-year cycle regarding their maximum and minimum number. Sunspots are regions where the sun's magnetic field breaches the surface and they usually appear in pairs. The reason they appear in pairs is because they are the visible part of magnetic loops, with each of the pair having opposite polarity (like the north and south of a magnet). Often one member of the pair is on the opposite side of the Sun's equator to the other. Sunspot maxima are the visible signs of the Sun's magnetic field flipping over. At each flip (every eleven years) the polarity of the sunspots reverses, so in fact it takes twenty-two years rather than eleven to complete the cycle. Counter-intuitively (because the sunspots are cooler than the rest of the Sun) a greater number of sunspots is associated with a more active Sun. The number of sunspots during different cycles can vary enormously and they may not appear at all for the peaks of some cycles. The reason for this is not clear. At the present time we appear to be experiencing the most sunspots (so most solar activity) for eight thousand years and up to now nobody knows why. The figure of eight thousand years comes from historical records and as such should be treated with caution, as the methods of recording have varied. However, before the modern era with its light pollution this count, on a hazy day, would have been pretty accurate. Tree ring counting has been used to back up these observations.

Although the sunspot cycle is associated with greater solar activity you shouldn't get too excited, as it seems that the sunspot cycle only makes a difference of 0.1% in the Sun's activity. To look at how this cycle has changed over the longer-term proxy data such as carbon 14 in tree rings can be used to measure the past activity of the Sun.

> **Carbon 14 in Tree Rings and Solar Activity**
>
> Carbon in the atmosphere (in the guise of carbon dioxide) has three naturally occurring isotopes, carbon 12, carbon 13 and carbon 14. The numbers refer to the total of number of protons and neutrons in each isotope and carbon 12 is by far the most common one. Carbon 14 does not occur normally but is produced when high-energy cosmic rays interact with nitrogen in the atmosphere. It is radioactive and decays with a half-life of about 5,700 years – a very useful half-life when looking over a period of several thousand years.
>
> When trees absorb carbon dioxide the carbon 14 to carbon 12 ratio at the time of incorporation is preserved in the annual rings of many trees. The annual rings vary in thickness with growing conditions at the time. Dendrochronologists (people who study tree rings to deduce the time signature) have built up a library of tree rings that record, among other things, climate for more than the last ten thousand years. The carbon 14: carbon 12 ratio in the atmosphere is influenced by the Sun's activity, because a more active Sun gives a more powerful Solar wind, which deflects cosmic rays away from the Earth, leading to a lower carbon 14: carbon 12 ratio.

Paradoxically there is a much firmer basis for estimates of the Sun's output over longer periods of time. It can be said, with a fair degree of confidence, that since the origin of the

Solar System (4.6 thousand million years ago) the Sun's output has increased by about 30%. The reason for this confidence is that the nuclear fusion reactions that power the Sun are well described by quantum physics. As the Sun continues to convert hydrogen into helium in its core, the density and temperature of the core increases, simply because helium has a higher mass and so, under these conditions, it has a higher density. The fusion process gradually moves outwards as the hydrogen is consumed and the ever-increasing temperature and density lead to a greater output of energy. Simple calculations indicate that the Sun will carry on increasing its output gradually by this mechanism for about five thousand million more years, when it will turn into a red giant (more of this later).

It is all very well looking at the Sun's output but how does this effect the energy flux (energy arriving over a specified time in a defined area) that impinges on the planets in the Solar System? Typically, emitted radiation from a source spreads out symmetrically in all directions and this is mostly true for the Sun, apart from the more directional nature of solar flares. Radiation that spreads out like this obeys a rule called the Inverse Square Law.

The Inverse Square Law

Common sense and experience tells us that the further away we are from a source of energy the less will be its effect. If you have a fire, for instance, you sit closer to it when you are cold and further away when you get too hot. The Inverse Square Law simply formalises this effect. It applies to most emitted radiation no matter what the wavelength and is a simple consequence of living in three-dimensional space. As the radiation spreads out in all directions, radiation emitted at one specific time occupies a spherical shell centred on the source of the radiation. If we are on the Earth at 1AU from the Sun and hold up a square metre of a detector of some kind we can measure the amount of energy falling on it (ignoring for the sake of clarity the Earth's atmosphere). If we now take the square metre detector to a position 2AU from the Sun and measure the radiation impinging on it we would find that it is a quarter of what we measured on the Earth. Similarly, if we were at 3AU from the Sun we would find that we measured the incoming radiation as being a ninth of that measured on the Earth and at 4AU it would be a sixteenth and so on. What we are seeing is the effect of the radiation propagating outwards on the surface of a sphere. The surface area of a sphere is $4\pi r^2$. Forget the rest, the r^2 is what matters because, in this case, r is the distance from the Sun and is the only thing changing in the equation. The consequence of this is that the energy emitted is being spread over a greater area as it

> *propagates away from the Sun, in such a manner that it is spread over four times the area when it is twice as far away, nine times the area when it is three times away etc.*

Looking at the Solar System Mars is 1.52AU from the Sun, which means that the radiation falling on a square metre of Mars from the Sun is 1/(1.52 x 1.52) the strength of that falling on the Earth (about 0.43) so Mars receives only 43% of the Sun's energy per unit area compared to the Earth. Now Mars is only one and a half times as far away from the Sun as the Earth so you can see how rapidly the flux from the Sun (or any radiating body) falls off with distance. Also remember that Mars is a smaller planet than the Earth, so actually receives much less total radiation on its surface. For Venus (I am using Mars and Venus as examples for reasons that will become clear in a little while) the distance from the Sun is 0.72AU. The same calculations show that it receives 1.93 times the radiation per unit area than that received by the Earth, i.e. it receives 93% more sunlight than the Earth. (As a matter of interest Mercury receives 658% more energy per unit area than the Earth.)

From earlier discussions it has been shown that Venus is too hot for life and Mars is too cold, while the Earth is just right (astronomers call this the Goldilocks effect, which shows that they do read books other than those about astronomy). The temperature of a planet's surface does not only depend upon the amount of sunlight received by it. Planets with a high albedo will reflect much of the incoming energy back into space. If there is an atmosphere this will absorb energy, especially when greenhouse gases are present, as well as distributing heat around the surface. In general planets with a high albedo will be cooler than those with a low one and planets with a lot of greenhouse gases (carbon dioxide, water vapour, methane and nitrous oxide) will be warmer than planets with less of these in their atmosphere. For life to be present on the surface of a planet there must be at least some liquid water present. Planets that orbit at the distance from the Sun (or another star) where liquid water can be present are said to be in the habitable zone.

The Habitable Zone

In the Solar System at present only the Earth has liquid water on its surface. Without its atmosphere temperatures on the Earth would be below the freezing point of water and it is the greenhouse effect that keeps the Earth habitable, so we can place the Earth in the habitable zone of the Sun.

Venus at present has no surface water and is very hot due to an overactive greenhouse effect, mainly due to carbon dioxide. From what is known of the formation of the Solar system it seems inevitable that any rocky planet will start out with a considerable amount of carbon dioxide in its atmosphere. Decreasing the amount of carbon dioxide in the atmosphere of

Venus to the levels found on the Earth would still not cool Venus enough for standing liquid water to exist, because of the amount of energy it receives from the Sun. So it seems that, at present, Venus is further in than the inner edge of the habitable zone around the Sun.

Mars has no surface water, that we know of, and has a very thin atmosphere (mainly because it is small and, almost certainly, has no tectonic activity to replace its lost atmosphere). It is theoretically possible that with a thicker atmosphere the greenhouse effect would raise the temperature of Mars so that it was within the habitable zone. So the outer edge of the present habitable zone of the Sun lies just beyond Mars.

As it is known that the energy output of the early Sun was only 70% of its present value, we need to look at how the Sun's habitable zone has changed with time and so find out which zone has been continuously habitable. Clearly as the temperature of the Sun was lower in its youth the habitable zone used to be closer in to the Sun and Venus, Earth and Mars would have received less energy. We know that liquid water has been present on the Earth more or less since its formation so our planet has always been in the habitable zone. Venus may also have been cool enough to have liquid water on its surface, although we cannot be sure. It would seem reasonable to put the inner edge of the Sun's early habitable zone at Venus' orbit. Intriguingly Mars shows signs of past surface water so, despite

> *receiving less energy in the past it must have had a thick enough atmosphere for a substantial greenhouse effect to occur, so it too was (just) inside the early habitable zone. As the Sun gets warmer the habitable zone is moving outwards, so Venus will never again be in the habitable zone and Earth may eventually leave it. Mars of course will be overtaken by the habitable zone, although its lack of atmosphere would still keep it too cold, and surface pressures too low, for liquid water to exist.*
>
> *There is a way around this for Mars. If we were to drop a few (A back of the envelope calculation suggests three or four) comets on to the Martian surface, this would produce an atmosphere full of water vapour, which is a good greenhouse gas. Regrettably this plan would upset quite a lot of people who want to look for indigenous life on Mars. On the other hand, warming Mars may well be good for any lurking Martian organisms. To sum up, the continuously habitable zone is a fairly narrow range of orbits, more or less centred on the Earth and possibly including Mars, but not including Venus.*

The rest of the Solar System is much too far away to be even considered for inclusion in the continuously habitable zone of the Sun in its present form. Although there may be life under icy surfaces, warmed by interior heat, this life won't amount to much as the habitat is marginal at best. However, in about five thousand million years, the Sun will become a red giant and the habitable zone could well reach the orbits of the gas giants and their moons.

An aspect of the Sun that is easy to take for granted is its stability. For over four and a half thousand million years the output of the Sun has gradually increased but, within that envelope, the short term activity of the Sun has almost certainly not varied by more than 1%. This is not to say that major events never happened. Solar flares tend to occur at the height of sunspot activity and give out very energetic x-rays and gamma rays. They are not conspicuous in the visible region as here they are swamped by the normal emissions at these wavelengths. Solar flares typically originate from an area less than 1% of the Sun's surface, so do not contribute much to the total emissions of the Sun. If a solar flare faces towards the Earth the event can be quite spectacular, giving rise to aurorae.

Aurorae
An aurora is a spectacular light show that occurs in the higher latitudes when solar flares impinge upon the Earth's atmosphere. They occur when highly energetic particles from the Sun encounter the Earth's magnetic field and are funnelled by this field down through the Earth's magnetic poles (which correspond fairly closely to the geographic poles). Aurorae are caused by the energy from solar flares ionising molecules in the

Earth's atmosphere, by stripping lots of electrons off. As the ions grab the electrons back the energy given out appears as sheets of light, outlining the magnetic field, that last hours or sometimes even days. In my experience aurorae are mainly green, red and purple, more obvious at night and at high latitudes (nearer the poles), but can sometimes be seen during the day with the sun behind you. What I find with aurorae is that they are beautiful and dramatic and really call for a sound track to enhance the experience. They are not that benign, however, as the associated magnetic effects can damage electrical systems, even causing black outs of power grids as happened in the USA and Canada a few years ago.

Another more violent event is a coronal mass ejection, sometimes called a magnetic storm. Here a magnetic disturbance from the Sun sends out a pulse of energetic particles, which can damage satellites and give a high dose of radiation to any passing astronauts. (If they are warned in time they are well advised to get some serious shielding between themselves and the incoming radiation.) Although these events are spectacular it is worthwhile remembering that they are well constrained within the less than 1% variation in the Sun's output figure quoted earlier. As to their effects on the Earth, these are transient and if we did not have power grids and people floating around in orbit they would hardly be noticed at all in the majority of places on the Earth. The major worrying aspect of these events is that if we were sending an expedition to Mars or elsewhere they could easily endanger the mission, a headache for Mission Control.

The gravitational effects of the Sun have a particular effect on maintaining life on Earth. As mentioned earlier plate tectonics is essential for recycling material on the Earth especially with respect to carbon dioxide. Along with the Moon the Sun has a tidal effect and the combination of these tides not only affects the surface water but also helps stir the mantle, which keeps plate tectonics running. Also the oceans are stirred by tides more effectively using two bodies rather than one, helping to move nutrients around and preventing stagnation.

Gravity is the second most noticeable aspect of the Sun, after it's radiation. As the Sun contains the vast majority of the Solar System's mass its gravity dominates everything else. Everything orbits around the Sun in one way or another. This is not strictly true, of course, as any two bodies will orbit about a point that depends upon their different masses. They will orbit about each other, about a point that is closer to the more massive body. In the case of the Solar System the next most massive planet is Jupiter, so Jupiter and the Sun orbit around a point in between them. Because of the large difference in mass between the Sun and Jupiter they do, in practice, orbit around a point inside the Sun. This also holds true for the lighter bodies in the Solar System, adding complications. These effects are, for most purposes, minimal, although they are large enough to be detected and are very useful when looking for stars that contain planets. Gravity appears to obey the Inverse Square Law and most observations support this, but there is a growing consensus that something might be wrong with this assumption.

> ### *The Pioneer Spacecraft and Gravity*
> *Pioneers 10 and 11 are spacecraft, which were launched in 1972 and 1973 to investigate the outer planets. After their mission was completed they continued on their way and will eventually leave the Solar System. These spacecraft continue to be monitored and a very strange effect was noticed. Both spacecraft have been cruising without power, so should only be affected by the Sun's gravity, however detailed information on their trajectories show that they seem to be experiencing a larger gravitational pull from the Sun than expected. This effect is in fact rather small, so most scientists consider it to be caused by some unknown heat source on the spacecraft. The niggling thing is that this effect is exactly the same for both spacecraft, which would imply the same unknown emissions. Although both Pioneers were built to the same specifications, it is difficult to see how a single, identical fault could be replicated so exactly, bearing in mind how long they have been in space. The*

> *alternative is that the laws of gravity that we know of are incorrect at large distances from the Sun. Perhaps these spacecraft have climbed out of the Sun's gravity well and reached interstellar space, where different gravity rules might hold sway in the absence of a massive body, though this sounds pretty unlikely too. To be honest this does not affect our understanding of the Solar System very much but, if this is not an artefact of the design of the Pioneers, it could cast light on dark mass and dark energy, which could be important on a galactic scale.*

One important aspect relating to the Sun's gravitational effect on the Solar System is the absence of any other star(s) close by. Many other stars in the galaxy are linked in binary systems, or even small groups. Another star, or stars, close to the Sun, would obviously complicate the gravitational environment of the Solar System. Complications on this scale would leave life forms feeling very unhappy, as habitable zones would come and go.

As the Sun moves through space it carries the rest of the Solar System with it (the Sun rotates around the Galaxy at 220 kilometres a second or 490,000 miles per hour. This is really fast!). Gravity is the reason that the planets are not left behind and we do not notice how fast we are moving. In its rotation round the Galaxy the Sun encounters random clouds of dust and sometimes comes closer to other stars, possibly stars that are going nova (exploding). These encounters are potentially lethal to life. Some mass extinctions on Earth could possibly have been helped along by such encounters. Altogether life on Earth has been extremely lucky, when you consider that a nearby supernova could literally sterilise the planet, or a close encounter with another star could pull the planets out of orbit.

This chapter has shown that the Sun is remarkably stable and has been for over four and a half thousand million years, despite a steady increase in its output. This stability has been kind to life on Earth, as it has given time for evolution to produce complex life. For the other planets around the habitable zone, Venus and Mars, it has not been so kind, indicating that the continuous habitable zone is in fact rather narrow. Obviously the Sun is not the only star that we know about, so next we look at how other stars compare with the Sun.

Where Are All The Aliens

Chapter 8

The Sun As A Star

To put the Solar System in context, it is useful to compare the Sun with other stars.

What is becoming less obvious as time goes by and light pollution increases is the huge number of stars that can be observed on a clear night. I am lucky to live in a small village with few street lights, many miles from the nearest city and so have a reasonable view of the night sky. I leave my curtains open at night and you would be surprised at the entertainment value in the sky at night. All the clichés usually used don't really do it justice. Apart from the obvious stars, on a clear night you can see a band of brightness across the sky which, when looked at with a reasonable pair of binoculars, or even better a decent telescope, resolves itself into even more stars.

The band of stars that we see is part of the Milky Way galaxy, which is where the Sun is situated. Fortunately we are out on a limb in the outer parts of the Galaxy (I will use the capitalised Galaxy when talking about the Milky Way, as it saves a bit of typing) so we have quite a good view of much of the rest of it. Actually this isn't just a fortunate coincidence as, for reasons described later, you would expect life bearing planets to be located away from the centre of any galaxy.

It is a sobering thought, but until the 1920s astronomers didn't realise that the universe contained anything more than the Milky Way. Now we know that this is just an ordinary representative of billions upon billions of galaxies! If observing from intergalactic space you would see the visible Galaxy as a central bulge, surrounded by at least three spirals in the shape of a disc. Of course, no-one has yet been into intergalactic space to view the Galaxy, but with modern, powerful telescopes we can see other galaxies, such as the Andromeda galaxy, the largest spiral galaxy in our local group. The patterns that emerge from the study of our Galaxy confirm its spiral shape. Galaxies come in a wide range of sizes, the Milky Way being about a hundred thousand light years across.

Light Years

A light year is simply the distance that light travels, in a vacuum, in one year. The speed of light in a vacuum is assumed to be a constant with good reason. The classic experiments by Michelson and Morley in 1881 and later supported the idea that the speed of light is a constant, though they were in fact looking to find the speed of the Earth through the ether, a fluid that was supposed to permeate the universe. Many later experiments seemed to confirm their results, though recently there have been some misgivings. Firstly, having aggregated their results (this is common practise) like most people who have carried out similar experiments, they may simply have thrown the baby out with the bath water and obscured other effects. More telling is the fact that we can only realistically measure the speed of light at present and it could easily have been different in the past or under extreme conditions. Recent results do imply that the speed of light may not in fact be a constant after all. However sticking to the general view, one light year is equivalent to about nine and a half million million kilometres, an extremely long distance, but very useful to astronomers. As an indicator the Sun is just over 8 light minutes away, the nearest star, Proxima Centauri is about 4.2 light years away (which is about average for our part of the Galaxy), the Galaxy is one hundred thousand light years across and the Andromeda galaxy, by way of contrast, is about 2.2 million light years away.

The visible matter in the Galaxy consists of stars, gas, dust and radiation. The Sun orbits the Galactic centre about once every two hundred and forty million years. There are also stars and groups of stars that do not move in the Galactic plane but orbit around the Galactic centre, much like Oort cloud objects orbit the Sun. Along with the visible matter there is also dark matter.

> **Dark Matter**
>
> *Dark matter is simply matter that we cannot detect, at least for the moment. We know that something like dark matter exists because the speed of rotation of galaxies is such that if they only contained what we can see they would fly apart. Dark matter can be detected, so far, only by its gravitational influence. It is estimated that dark matter has a mass ten times that of the mass of the stars in the Galaxy. This seems to hold for other galaxies as well. The nature of dark matter is unclear. It cannot be totally accounted for by normal matter that we cannot see, because there is so much of it we would be constantly aware of it blocking our view. Two different approaches are that dark matter consists of a lot of massive sub-atomic particles (called WIMPS or weakly interacting massive particles) that haven't been detected yet, or that there is something wrong with our understanding of gravity on a large scale and there is no dark matter after all. Although I am usually willing to argue for one side or another I can't make my mind up about dark matter. Physicists love to invent new particles to explain everything, so are not reliable witnesses, but equally well a rewriting of gravity seems somewhat arbitrary.*

Another group of objects that exist in the Galaxy are black holes.

> **Supernovae, Neutron Stars and Black Holes**
>
> *A young star starts off by using hydrogen as a fuel, converting it to helium. As time goes on and the hydrogen is less abundant then helium begins to be used, forming heavier elements, which in turn undergo fusion to form even heavier ones. In fact this is the major way elements up to iron are formed. Heavier elements than iron do not support further fusion reactions. When a large star, a few times larger than the Sun, has run out of material to keep atomic fusion going then the core of the star collapses very rapidly. There is a rebound effect and the outer shell of the collapsing star, fuelled by the gravitational energy of this shell collapsing onto the core, explodes and for a brief time shines brighter than a whole galaxy. This is a supernova. (There are other ways of getting a supernova but this is the most common). The enormous energies involved allow elements of greater atomic mass than iron to be produced, some of which are necessary for life as we know it.*
>
> *What remains of the core continues to collapse until all the electrons and protons are squashed together and form neutrons. At this point the atomic forces between the neutrons <u>may</u> stop further gravitational collapse of the star. Needless to say this is a neutron star.*
>
> *Neutron stars are typically about 10km across and have a mass*

of a few Suns. However, if the core of the neutron star is greater than about five solar masses atomic forces cannot overcome gravity and the star collapses further to a singularity. A singularity is a point that has a high mass and no volume. Of course this makes little sense in the world of normal experience, but there is a way to describe this state using quantum physics. In most models of the universe this singularity sits at the centre of a black hole.

The reason that a black hole is called this is because at a particular distance from the singularity, depending on its mass, gravity is so strong that even light cannot escape directly - although modern theories have found a way in which light and other particles might leak from a black hole. The boundary at which light cannot escape from a black hole is called the event horizon. A black hole is fully described by the three parameters of mass, spin and electric charge. Obviously an isolated black hole is difficult to observe, although a few isolated bodies are thought to be black holes forming. The main observational evidence for black holes come from the interactions with their surrounding environment. Binary stars (two stars orbiting about each other; there are also multi star systems with necessarily complicated orbits) account for about half the stars in the Galaxy, so you would expect to find some black holes orbiting another star and this is indeed the case. When in orbit about another star there is the opportunity for a black hole to pull material from the star and this does happen. In such a binary system material flowing from the companion star spirals in towards the black hole emitting radiation, mostly in the visual and ultraviolet wave lengths, but also as X-rays when the material encounters the event horizon, and these can be detected from Earth.

In fact this mechanism can produce black holes as, if a neutron star is in orbit about a sufficiently large star then material will be accreted onto it and its mass therefore increases until it collapses into a black hole. Most of the known black holes in the Galaxy that have so far been detected are in binary systems. If they have sufficient material falling into them, black holes can become enormous. The centre of the Galaxy contains such a supermassive black hole, with a mass about two and a half million times that of the Sun in a very small volume. Some other galaxies have a central black hole with two hundred times this mass.

One interesting theory is that in the early stages of the Universe an enormous amount of mini black holes would have formed, but as no one has ever observed one they remain in the world of theory. In the future, with a big enough nuclear collider, it may even be possible to make one, although a mini-black hole floating around the Earth might be a tad unsafe.

As all evidence indicates that life can only originate on a planetary body circling a star this is the obvious place to look for life. There are about one hundred thousand million stars in the Milky Way galaxy (estimates vary and this is on the low side, but the discrepancy relates to how you count very small stars called brown dwarfs, which in terms of supporting life are unimportant).

Although it isn't obvious, because we can only see them as point sources, stars have different colours i.e. they emit light (and other electromagnetic radiation) more or less strongly in certain parts of the spectrum. A classification of stars based on their visible spectra gives useful information, especially their surface temperature. This classification is possible because the hotter a star is the higher proportion of the energy is emitted at the blue end of the spectrum.

When you look at the sky virtually all the stars that you can see look white. This is an artefact of human vision as in dim conditions we see in black, white and shades of grey. The stars are simply not bright enough for the human eye to detect their colour as not enough photons reach the eye to trigger our colour detection system. Even with a moderate telescope it is difficult to pick out these colours, but spectrographic methods can detect them to a high degree of accuracy.

What comes out of this is the Harvard Spectral Classification, where stars are assigned a letter from the sequence O, B, A, F, G, K, M with the hottest stars being O and the coolest stars M. (The reason for selecting these letters is purely historical, having to do with the detailed spectra of stars - interesting it may be but really not worth pursuing any further). Within this sequence stars are subdivided further. For stars in classes B to M each star also has a number 0 to 9, with 0 being the hottest and 9 the coolest star for a particular letter. Rather eccentrically, stars of classification O are subdivided with the numbers 5, 6, 7, 8, 9, 9.5, with 5 being the hottest. (A lot of astronomy is filled with these quirks). The following table lists the middle of the range spectral type of stars against their surface temperature (except for O which is given as its highest temperature, simply because at such high temperatures the upper limit is difficult to define).

Spectral type	Temperature in degrees Kelvin
O5	40,000
B5	15,500
A5	8500
F5	6600
G5	5500
K5	4100
M5	2800

This list is also a list of abundances. There are very few of the really hot stars, somewhat more of the intermediate stars and even more of the coolest ones. The spectral classification of the Sun is G2, so it isn't as exciting as a really hot class O star, nor is it as boring as a class M one.

Since the stellar types were first introduced another one, type L, has been added for stars with a lower temperature than M class stars. These are called brown dwarfs and radiate in the infra-red part of the spectrum. The existence of brown dwarfs is one of the reasons that the estimated number of stars in the galaxy is disputed, as they are of such a low luminosity. They are likely to be as numerous as all the other stars combined and typically have about 8% the mass of the Sun. Brown dwarfs undergo nuclear fusion but this doesn't extend to the surface, so they are on the boundary between a large planet and a small star.

The overall behaviour and evolution of stars, except where they are directly interacting with other stars, is almost entirely dependent on their mass. When first formed all stars have relatively low brightness and temperature, after that things get a little more complicated. As fusion really gets underway the differences begin to show. At this stage stars of a lower mass than the Sun burn less brightly and at lower temperatures than the Sun, then they settle down and slowly increase their temperature and brightness. After a very long time (there is a chart later that tells you how long) they start to run out of hydrogen fuel and lose their brightness but increase their temperature and become what are called white dwarfs.

The period that a star spends slowly increasing both its brightness and temperature is called the Main Sequence lifetime of that star. All stars spend a certain amount of time on the Main Sequence, although very low and very high mass stars, for different reasons, only spend a brief time there. Interestingly, the age of the Universe is only about 14 thousand million years, which means that no stars much smaller than the Sun have had time to leave the Main Sequence, so any white dwarfs out there must be from Sun-sized stars or bigger.

Stars of the same mass as the Sun spend a shorter time on the Main Sequence than smaller stars, although ten thousand million years is not to be sniffed at. Whilst on the Main Sequence, sun-like stars are both brighter and hotter than lower mass stars. When their hydrogen reaches a low level these stars become cooler, brighter and much bigger, so they are called red giants (the red colour indicates that they are cooler). When most of their nuclear fuel is consumed sun-like stars become white dwarfs.

Stars of higher mass glow more brightly and at a higher temperature than the Sun. Their Main Sequence lifetime is very short, relatively speaking. At the end of their Main Sequence lifetime they turn into supergiants, which are brighter than red giants, but have a similar temperature, so unsurprisingly they are also red. The final fate of supergiants is spectacular, as they blow up, becoming supernovae and leaving behind a neutron star.

Mass compared to the Sun	Temperature/K	Approximate main sequence lifetime in years
25	35000	seven million
15	28000	ten million
9	23000	twenty million
5	17000	seventy million
3	11000	two hundred million
1.5	7000	two thousand million
1	6000	ten thousand million
0.75	5000	thirty thousand million
0.5	3800	two hundred thousand million

The importance of the Main Sequence lifetime of a star is that, while on the Main Sequence, any planets around the star will not be subject to sudden large changes. Once off the Main

Sequence sudden changes do occur which will, almost certainly, destroy life on any planets orbiting the star.

For stars greater than about 1.5 times that of the Sun the Main Sequence lifetime is not long enough for life to originate and evolve far enough to develop intelligence. Bear in mind that the Sun is four thousand six hundred million years old, so is about halfway through its Main Sequence lifetime and yet the Earth has only just arrived at the point where intelligence has evolved. There is of course the possibility that intelligence may develop later or earlier than on the Earth. Obviously later development precludes life on planets around a star heavier than on the Sun because of the shorter period of stability. If you look at a star with 1.5 times the mass of the Sun, its Main Sequence lifetime is only a fifth that of the Sun, so is much less likely to have a life bearing planet around it – and the length of time on the main sequence diminishes very rapidly with higher mass stars. Earlier development may be possible, although life does seem to have developed on Earth due to a lot of unlikely events happening in sequence.

For stars much less massive than the Sun their Main Sequence lifetime is very long, so there is no problem with instability to upset the development of life. In fact there are so many of these smaller stars that they seem to be the main candidates for life bearing planets. There is a catch however. Stars of less mass also have considerably lower surface temperatures. The lower temperatures of such stars means that the habitable zone must be much closer in and so any planet orbiting them would have to be in a much closer orbit to stand a chance of supporting life. For example, if a star has half the Sun's mass its habitable zone will be similar to Mercury's orbit. The trouble with a planet orbiting too close to the parent star is that it becomes tidally locked into synchronous rotation, so it always has the same face towards the star. (This would be true for Mercury, but this planet has probably been hit by another body being pulled into the Sun, a not unexpected consequence of a close orbit.) In these circumstances one hemisphere would be very hot and the other very cold - enough to preclude life altogether.

It has been suggested that this problem could be overcome if the planet had a dense atmosphere, in which case atmospheric circulation could distribute the heat from the hot to the cold side. On closer inspection though, this seemingly viable situation would not last long. On the hot side of such a planet the kinetic energy of the molecules would mean rapid loss of gases from the top of the atmosphere; on the cold side atmospheric gases would condense and be lost from the atmosphere. Besides this, while the atmosphere lasted there would be incredibly strong winds blowing all the time, driven by the temperature difference, generating massive erosion. Another idea is that life could develop in the area between the hot and cold hemispheres, but this would be a marginal – and very windy - existence at best. Also smaller stars would have formed from smaller dust clouds so would be unlikely to have large enough planets for life to originate.

The conclusion then is that complex life can only develop on planets circling a star similar to the Sun (i.e. greater than a half, but less than one and a half times the Sun's mass). This is, at least on the surface, a reasonable option as about 10% of stars fall into this category, which gives us a possibility of ten thousand million such stars having life bearing planets to look for in the Galaxy. Of course, on average only half of the Sun-like stars will be as old or older than the Sun and only these can sensibly be considered to have developed intelligent life.

As mentioned earlier, within the Galaxy about half the stars are in a binary or multiple systems. Planets may form within such systems, especially if the stars are far apart. In fact some, a very few, have already been detected. Planets in a binary system can orbit one star fairly close in when the stars are far apart or orbit both stars when the stars are close together. Stellar systems are formed out of a rotating disc of dust and gas that can form a central star with orbiting planets due to accretion of the gas and dust. In the case of a binary system with widely separated stars it is likely that they each formed from their own discs, otherwise, if there was only one disc, there would not have been enough material to form the star on the edge of the disc. Under this scenario each star would have its own circle of planets (providing there was enough material to form them) and inevitably there would be interactions between the planets circling one star and those circling the other, as well as interactions with the other star. This is neither a stable nor predictable situation. If such planets exist, the orbit would become increasingly elliptical and hence unstable and not suitable for life.

A close binary system can form when a large cloud of dust and gas contracts to form twin centres. Planets orbiting a close binary will necessarily be subjected to gravitational attraction from both stars, which will fluctuate depending on its position on its orbit. The orbit itself would be erratic and so if a planet were at the right distance from a sun to be in the habitable zone sometimes, it would drift in and out of the zone during one revolution, clearly not a recipe for a life-bearing planet.

The conclusion must be drawn that no advanced life is likely to be found on the planets in a binary (or any multiple) system. At this point it is worth considering where the stars and planets ultimately came from. The best theory we have at the moment to explain the origin of the Universe is the Big Bang theory.

> ### *The Big Bang*
> *All the evidence points to the Universe originating from a singularity fourteen thousand million years ago. This singularity flew apart creating its own space and time and this expansion is not only continuing today but seems to be accelerating. The details are really for another book as it is not a simple scenario. For the moment what is important is that all the matter that we see today originated in the Big Bang, but with a difference. The matter created by the Big Bang and its aftermath was almost entirely made up of hydrogen and helium with miniscule amounts of lithium and beryllium.*

Although lacking in heavy elements the first stars would seem to have been very large and hence evolved very fast. Nuclear fusion in these stars created elements of higher atomic mass than hydrogen and helium. Because they were massive, these stars would have ended their lives as supernovae, creating even heavier elements and exploding them out through space. We have no direct evidence yet of these stars, but they must have existed otherwise we would not be here to tell the tale. (The Hubble Space Telescope has recently found galaxies very far away; as the light must have taken a very long time to get here this implies they are from the early Universe. These contain very bright blue stars, which may be this evidence.) As these heavier elements were flung throughout space they became incorporated into a second generation of stars. This second generation then synthesised more of the heavier elements and also spread them through space and so it continued. The distribution of these heavier elements depends on where the parent stars were when they exploded and this has led to a patchy distribution of these elements through space, although still mostly contained in galaxies. When we look now there are stars and interstellar clouds that vary in their metallicity (a factor mostly ignored by astrobiologists) and thus in their ability to form rocky planets.

Metallicity

Metallicity is just a way of describing the amount of elements heavier than hydrogen and helium in the object being observed. The earliest stars that were formed after the Big Bang would have contained only the elements that were formed at the birth of the Universe. As no elements of higher atomic mass than beryllium existed at this time the metallicity of these stars would simply be the percentage (by mass) of lithium and beryllium contained within them. This is fairly reliably estimated to be one ten millionth of a percent. As a comparison the Sun has a metallicity of about 2%, which is reasonably high as these things go. The metallicity of stars will obviously increase with time as nuclear synthesis has only transmuted a small fraction

> *of the available hydrogen and helium. Elements with an atomic mass up to iron (atomic weight about 56) are produced in large stars by fusion reactions. Elements heavier than this cannot be formed by fusion as it takes too much energy to create them by this route. These higher elements are produced by a process called neutron capture, which, involves high-energy neutrons being captured by heavy elements, mostly in supernova explosions.*

In terms of the formation of planets metallicity is very important. Planets form by the clumping together of the leftover bits of the stellar system, after the star's formation. This largely depends on mass and density and so is a lot easier when heavier elements are present. Stars with lower metallicity than the Sun have difficulty forming planets and many seem not to have any at all. This may well be because a dearth of heavy elements means a slow coalescence of the small bodies that lead to planets, so that at the T-Tauri phase virtually everything developing in the nebula would have been swept away. At the other end of the scale, stars with a greater metallicity than the Sun primarily form giant planets. Interestingly this implies that as the Galaxy evolves there will be a move towards stars having ever-larger planets.

The Sun is out on the edge of the galaxy and this is one of the reasons that it has habitable planets. Closer in to the galactic centre the stars are much more closely packed and so gravitational disturbances would be more likely to affect any planetary system. Probably of more importance is that closer to the Galactic centre there are many more times the number of supernovae than further out. A supernova close to a stellar system, spewing out lots of harmful radiation, would severely damage any life bearing planets, so it is not a good idea to be near one. It seems that the Solar System has, purely by chance (I think), not had a supernova close by for so long that it is bucking the odds. This doesn't mean that this situation will last forever as supernovae are unpredictable, or at least we can't predict them yet.

During its orbit around the Galaxy the Sun, and of course its attendant planets, moves through areas containing varying amounts of dust. There is some evidence that higher levels of interstellar dust may have contributed to extinctions on the Earth, a good reason for the Sun to avoid the inner regions of the Galaxy with its much higher density of dust clouds.

In our search for possible life-bearing planets then, it seems we must limit our search to the outer regions of the Galaxy (and any other galaxy too). The turbulent, dangerous conditions in the more crowded regions would regularly wipe out any life that managed to get started there.

So in this chapter we have seen that only a small percentage of stars can possibly host planets on which intelligent life can develop. These stars must be more or less Sun sized, not in binary or multiple systems and in the outer regions of their parent galaxies.

Of course these strict criteria need not be applied in the search for simple extraterrestrial life. As we have seen on the Earth life is pretty ubiquitous and I believe is likely to develop whenever a planet spends a while in a habitable zone. However, long-term stability is necessary for more complicated life forms to evolve. Next we look at what is known at present about planets orbiting other stars.

Where Are All The Aliens

Chapter 9

Planets Beyond The Solar System

Before 1992 it would not be possible to write about planets beyond the Solar System, otherwise known as exoplanets, as that was the time the first two were discovered. These first exoplanets set the scene for other studies as they are simply bizarre; both these planets were discovered orbiting the same pulsar and by all common sense should not have been there.

> *Pulsars*
>
> *Pulsars, like other neutron stars, are the remnants of supernovae that rotate very rapidly, averaging about a hundred times a second (the fastest rotating pulsar yet found rotates 642 times a second). They have strong magnetic fields that are not aligned with the rotation axis of the star and these strong, rotating magnetic fields produce a directional beam of radio waves. As the magnetic axis is offset from the rotation axis we see the radio waves strongly when the magnetic axis points*

> *towards us and not at all when it points away from us so the pulsar appears to pulse at a rate which is the same as the rotational period of the parent star.*

The huge energies involved when a star explodes as a supernova should have vaporised anything in the vicinity and so it was an enormous surprise to find planets orbiting a pulsar. Astronomers have come up with three possible scenarios to explain their existence. Firstly, they may have been captured after the supernova explosion, though this seems unlikely, bearing in mind the huge distances between stars and the fact that there are two planets involved. They may be the remnants of gas giants that survived the supernova explosion with rocky cores left intact or lastly they may have been formed from in-falling material that did not entirely escape the supernova's gravity. Of all these scenarios I consider the last option to be the most likely.

Whatever mechanism holds it is quite clear that neither of these planets could possibly hold life. After all, a pulsar has an intense magnetic field, which is rapidly rotating and gives off lots of damaging radiation. It is ironic that when the first pulsar was discovered in 1967 it was given the designation LGM (after 'little green men') as such a regular pulse seemed impossible for a naturally occurring object.

In 1995 the first planet orbiting a main sequence star was found and more discoveries followed. To date (2007) about one hundred and seventy exoplanets have been found and their number is increasing rapidly. This rate should increase when the proposed orbital telescopes are deployed (these unfortunately keep getting delayed). So far, to everybody's surprise, all the exoplanets found seem to be very different to our experience of planets in our Solar System, even taking into consideration that only large planets can be detected with the telescopic power at our disposal. In fact most of them turn out to be "hot Jupiters".

Hot Jupiters

'Hot Jupiters' is a term invented to describe planets that are of comparative size to Jupiter but orbit close to their parent star and so, because of this proximity, have a high temperature. Most hot Jupiters so far found have a minimum mass in excess of that of Jupiter. (Minimum mass has to be used because the methods used to detect these planets are affected by the angle that their orbits present themselves when observed from the Earth so you can only estimate their mass). The maximum mass of a hot Jupiter would be about thirteen times that of Jupiter as, at this mass, fusion of deuterium (a rare isotope of hydrogen with an extra neutron, which is a prime candidate for use in

> *fusion reactors or starship engines) can occur with a high degree of efficiency and such objects are best described as stars. The fact that so many hot Jupiters have been found was a total surprise. They really shouldn't be where they are, as their parent star would have ripped them apart during its T-Tauri phase and they certainly would not have had enough time or materials to form after the T-Tauri phase. Many hot Jupiters are so close to their stars, often closer than Mercury is to the Sun, that their orbital period is expressed in days. This is very different from Jupiter, which has an orbital period of 11.86 years.*

The most accepted theory for the origin of hot Jupiters is that they formed further out in their star system and so would be less affected by their star's youthful tantrums. In this scenario these planets would migrate inwards due to friction with the stellar nebula in which they are embedded. This has clearly not happened with Jupiter (although some inward migration is generally accepted), which begs the question - why not? One possible explanation is that stellar systems with hot Jupiters tend to have a higher metallicity than our Solar System. A high metallicity implies more material to build planets and more rapid growth of planets. In such a system many more opportunities arise for formation of rocky planets and other bodies such as asteroids and comets. Looked at this way a Jupiter (or a super Jupiter) would have a high rate of collisions with other bodies, causing it to lose angular momentum and so spiral in towards its star, picking up mass as it went. This is, of course, a recipe for disaster, as if you are a rocky planet in the habitable zone you are likely to be swept up by the in-falling giant planet.

The high metallicity observed for these stars does not necessarily prove that the original nebula had a high metallicity. For example if a large planet had fallen into one of these stars there would be a signature from the destroyed planet, characterised by heavy elements in the outer regions of the star, which would show up as a high metallicity. A temporary phenomenon, but it would last for a few million years. This is probably not a common mechanism however, as it does imply an unrealistic amount of planets crashing into stars recently.

There have been computer simulations undertaken by astronomers at the Open University in the UK, who show that it is possible to create models of stellar systems, containing hot Jupiters, where you can slot an Earth sized body into a couple of stable points in the stellar system (called Lagrange points) so they avoid being consumed. Much as I dislike to disagree with the Open University astronomy department (not least because the Open University was where I obtained my degree and I am a firm supporter of the University), it seems to me that these points of stability will vary as the hot Jupiter spiralled in and as such are only stable after the event. During such an inward spiral there would be nowhere to hide and so no surviving rocky planets, as they will either be consumed by the hot Jupiter, thrown into the star or thrown out in a highly eccentric orbit. None of these options are likely to leave a life-bearing planet.

The fate of hot Jupiters doesn't seem to have been addressed at all sensibly. Firstly, it is not at all clear what process prevents these planets from falling into their stars. By the time they come into a close orbit of their parent star, most of the original stellar disc material would have been subsumed by the planets and the star and so friction from this source would have been mostly eliminated. The star however, particularly if it has a high metallicity, will be throwing out material via the stellar wind and, as the hot Jupiter is close in, this will further degrade its orbit. (This happens less for rocky planets as they have a higher density, so are much smaller than a gas giant of the same mass and so experience less drag). On the other hand a strong stellar wind and/or massive flares could act to push hot Jupiters away from the star.

The conclusion I come to is that hot Jupiters are, in the context of Galactic time, probably transitory phenomena, which is bad news for life as there will be many stellar systems where we will not be able to observe hot Jupiters because they have fallen into their parent star, taking any rocky planets with them.

As there are such a high number of observed hot Jupiters there are probably many more systems that used to contain them where this has already happened. This bodes ill for intelligent life as, if this is happening, inhabitable planets are probably few and far between in the Galaxy and the Solar System is a very rare occurrence.

One possible habitat for life could be on large moons around hot Jupiters as, by analogy with the Solar System's gas giants, there should be a lot of moons around such planets. In fact such moons are likely to be larger than similar moons in the Solar System because the environment in which they formed often had a higher metallicity. Europan analogues with icy surfaces and subsurface oceans would provide possible habitats, but marginal for life, as is the case for Europa. As there ought to be moons of similar mass to the Earth, there is the intriguing possibility of moons with a decent atmosphere. Such moons could derive a good part of their warmth from tidal forces. There is however a problem: To generate enough heat this way the moon must be subjected to much larger tidal forces than the moons of Jupiter. The surface of such moons would be incredibly active volcanically so any life would live in a very unstable environment indeed. This may be okay for simple life, but it is difficult to see how complex life could evolve, though it cannot be entirely ruled out.

Unfortunately, for life, the very nature of hot Jupiters mitigates against it. Hot Jupiters are unstable; they definitely formed much further out from their stars than the orbit in which they are observed now. The inward migration of these planets (and obviously their moons) mean that conditions on the moons would have changed dramatically over their lifetimes and it is not known if their inward migration has stopped. Probably the biggest hurdle to life on such moons however is the proximity of their stars. Most hot Jupiters are found very close to their stars, much too close and therefore too hot to be in the habitable zone.

There has to date (April 2007) been one planet identified that may be a candidate for life. This planet orbits a red dwarf star, called Gliese 581, 20.5 light years away. The planet lies within the habitable zone of its star and has a mass about five times that of the Sun. From its position in the habitable zone it would be reasonable to expect that it has liquid water and its size suggests, to many people, a rocky planet, although none of this is confirmed. Of course we do not have a similar planet in the Solar System to compare it with, as all the local planets are too big or too small. Interestingly, with this mass and assuming a moderate temperature, this planet should have an escape velocity that means its atmosphere is rich in hydrogen and helium. Of course we should not forget that it will also have a large carbon dioxide presence in its atmosphere with the accompanying greenhouse warming. I consider the jury still out on this being a rocky planet as the only evidence is its mass. As this planet will have retained a lot larger and denser atmosphere it could well be an entirely new object somewhere between a rocky planet and a gas giant. Of course being close to a red dwarf it is likely to be in a synchronous orbit with its star. Still this is our best bet so far for an exoplanet with life although the synchronous orbit would limit this to primitive organisms as mentioned before.

To pull back a bit, one of the main reasons that hot Jupiters are by far the most detected planets is a result of two possible factors. Systems with hot Jupiters may be the most common planetary system around (which the evidence seems to support) and/or the methods that we use for detecting exoplanets are especially good at detecting hot Jupiters (which is also definitely true). So far little has been said about the methods used to detect exoplanets so it is time to consider these. The most successful method for detecting exoplanets is by using Doppler Spectroscopy.

Doppler Spectroscopy

The Doppler effect is experienced quite commonly by anyone who takes a train. As you are standing on the platform and hear the train coming it has a high pitch (think frequency), as the train passes you and disappears into the distance the pitch drops. What you are observing is best looked at from the point of view of the wavelength of the sound. As the train is coming towards you it emits a sound with a particular wavelength; if you imagine looking at the peaks in the sound waves, the first one is emitted and then, before the next peak can be emitted, the train has moved towards you. The next peak therefore is emitted when the train is a bit closer, so you hear it slightly before you would if the train was stationary and it appears to have a higher frequency. After the train has passed you the reverse happens as in this case the second peak is emitted slightly further away than if the train was stationary, so it appears to have a lower frequency. This observation applies to any waveform and light also behaves accordingly.

Planets in a stellar system orbit their star but equally the star orbits the planet, or to put it correctly they both orbit around a point between them. If the orbit of a planet about its star is edgewise on for us viewing it, then the star will either be approaching us or receding from us as it orbits around the centre of gravity of the system. This means that light emitted from the star will seem to have a shorter wavelength when it is approaching and the opposite when it is moving away. Measuring this difference in wavelength gives you the orbital period of the planet as well as its mass fairly accurately. The mass calculated will be a minimum, as most stellar systems will

> *not be observed edgewise.*
>
> *Hot Jupiters are massive and close in to their star, so have a relatively large Doppler Effect (the closer the planet is to the star the greater the Doppler Effect as the orbital period is shorter and gravitational effects on the star greater). An Earth like body would have too small an effect to be detectable in most cases, so there is an inbuilt bias towards large planets in close orbit about their star. This is the reason we observe more hot Jupiters than we would expect on average.*

The most obvious other method of detection is where the planet passes in front of its star (i.e. occults it), which will cause a dip in the brightness of the star at periodic intervals as the planet passes between it and the Earth. This is called the transit method and a few planets have been found this way. It is a very limited method, as the stellar system being observed would have to have its ecliptic (orbital plane) virtually edge on to us. This method also favours large planets in a close orbit of their star, as such planets will cut out more of the star's light and will do so more frequently than any small planets further out. However, we can measure such dips in brightness to a fairly high degree of accuracy and so could detect planets smaller than those detected using the Doppler method.

There are other methods of exoplanet detection. A very interesting method is called gravitational lensing and deserves a little box of its own to explain it.

> **Gravitational Lensing**
>
> *This concept started with Einstein, who calculated that light would be bent if it passed through a strong gravitational field; this could be as a result of gravity bending the space around a star. This theory was first tested during a solar eclipse, when a background star appeared to be in a different place than would be expected using Newtonian physics. If a star, or planet, sits between the Earth and a distant object then light waves will be bent around it and under the right conditions the background object will be magnified as the star or planet acts as a lens, very much as a magnifying glass does. Admittedly a huge magnifying glass would be required, but the principle is the same. What you actually observe, in the best-case scenario, is a ring around the star or planet called, rather unimaginatively, an Einstein ring. The size of the ring relates to the lensing object (the star or planet in this case) and not the background object. If a planet is the lensing object its mass and distance can be calculated quite easily.*

Gravitational lensing has located a few exoplanet candidates, which have yet to be confirmed by other methods (although by the time you read this they will almost certainly have been confirmed). This method favours large planets once again but can detect such a planet at a great distance, though in fact the Doppler effect and transit methods are also independent of distance.

Astrometric measurement is simple to explain but in practice takes a lot of observational time. Stars have a movement relative to the Solar System, which means they don't just sit there in space - they zap along, sometimes at very high velocities. (Remember also that the Sun has a velocity of 220 kilometres a second with respect to the Galaxy.) This movement is essentially in a straight line as the orbit about the Galactic centre has a very large radius. An orbiting body (another star or a planet) attracts the star being observed and they orbit about their combined centre of gravity giving a regular wiggle around the straight line that can be detected. The mass and orbital period of the body can be deduced from analysing this wiggle. Astrometric measurements are much more likely to be used to detect binary stars than planets and have been successful in such cases, but this method has also picked up some hints of potential planets. Once again massive close-in planets are more easily detected.

Finally, although it probably should have come first, is direct observation. This tends to be described as detecting a firefly against a car headlight, but there are approaches that reduce the difficulty. Firstly, telescopes can be equipped with a disc to cover the star so the planet is not swamped by the glare; this method is commonly used and will be used by a new batch of orbiting telescopes with improved methods of covering the star. Secondly, planets emit their energy (actually, of course, reflected energy) towards the red end of the spectrum whilst Sun like stars are bluer; filtering out the blue part of the spectrum allows the planet to be observed more clearly. This method has not detected any planets, so far. Unlike the other methods direct measurement favours planets in orbits further out from their stars.

So far I have only looked at stars in isolation (other than binary systems) but stars are certainly not born in isolation. The Milky Way Galaxy is a relatively large spiral galaxy as these things go. There is little point looking at other galaxies in the pursuit of life, as they are unimaginably far away, however it is useful to put our Galaxy in context. There are, as a minimum educated guess, a hundred thousand million galaxies in the universe. Just like stars the smaller galaxies are more numerous and more difficult to detect, so can easily be missed. This means the true number of galaxies is almost certainly much greater than the figure quoted.

Types of galaxies fall into a natural classification based on their overall shape and size. Elliptical galaxies are the most common and have no central bulge but are brightest in their centre, with this brightness decreasing towards their outer edge. Overall, elliptical galaxies are mostly featureless and do not contain dust except for some very rare exceptions and these rare exceptions can, almost always, be associated with recent collisions with spiral galaxies. Star formation in virtually all ellipticals stopped a long time ago, so they are mostly made up of older stars. This is because they ran out of the gas and dust from which stars are formed. Now, heavier elements come from supernovae explosions and so as time goes by more and more heavy elements enrich the gas and dust and younger stars include this stuff in their formation. Older stars have a lower metallicity than younger stars, because at the time of their formation the gas and dust that they derived from had a lower metallicity. As we have seen, life cannot occur in a low metallicity environment, not least because rocky planets won't form. Elliptical galaxies seem to be the final result of collisions between galaxies, as if you go further back in time (which is what you are doing when you look further into space) there were less ellipticals present.

The next category of galaxies is lenticular; these have a central bulge and a disc but no visible spiral arms. They also seem to lack dust and are made up from older low metallicity stars.

Spiral galaxies like the Milky Way have a central bulge and a number of spiral arms; importantly the spiral arms have copious amounts of dust and many young stars occupying the spiral arms.

Irregular galaxies are the anarchists amongst galaxies. They have no discernable form, little dust and an apparently random distribution of stars; they are also generally small.

Galaxies tend to come in clusters and the Milky Way is no exception. In fact some of the small companion galaxies are known to be colliding with it. M31, the Andromeda galaxy, is the largest one of the local group to which our Galaxy belongs and, like the Milky Way, is a spiral galaxy but just a little bit larger. Observation of the Andromeda galaxy originally led astronomers to recognise that the Milky Way was a spiral galaxy.

Most galaxies seem to have a massive black hole at their centre. In the Milky Way Galaxy the central black hole is over two and a half million times the mass of the Sun, which seems pretty large, but other, more active, galaxies can have central black holes fifty times as big. Many of these are incredibly active, sucking in gas, dust and stars and emitting copious amounts of energy. If someone ever offers you a trip to one of these active galaxies, turn it down. It is not a place that life would find attractive. The Milky Way's black hole is not very active at present but shows signs that it once was and probably will be again.

Galaxies are a bit like stars when you consider whether they are inhabitable. The galaxies that lack dust have older stars with low Metallicity so are not a good place to look for planets of any sort. The spirals with their dust, gas and high metallicity are good places to look for planets. Again like stars, there are places of high potential for planet bearing stars within spirals as well as places that have low metallicity, where planetary formation is restricted. In a spiral galaxy the low metallicity stars lurk in the central bulge and halo objects. Halo objects represent a small percentage of the stars in the galaxy and occupy a slightly flattened sphere with a diameter the same as that of the Galaxy. They include, globular clusters that, as the name suggests, are globe shaped clusters of several thousand stars, these stars are all very old as the gas and dust in such clusters were used up a long time ago, they also have low metallicity so are unlikely to have planets. There are also a few isolated stars in the halo that have been thrown out of the Galactic Disc; these are also generally very old. The habitable (in the sense that they might have habitable stellar systems) areas of these galaxies are in the spiral arms. As the spiral arms of the Galaxy are the main sites of star formation, they need a little explanation.

It is tempting to assume the spiral arms are a permanent fixture, made up of stars, dust and gas. Nevertheless they are much more transient than that and the stars, dust and gas can be seen entering and leaving these arms. In fact, if they were permanent they would quickly wind up and disappear, as the galaxy rotates. No satisfactory theory has yet been able to account for all the attributes of the spiral arms but the best bet seems to be Density Wave Theory. This theory suggests that there are density waves orbiting the Galactic centre which naturally form a spiral pattern. These waves are not the same as the dust, gas and stars but appear to be a gravitational phenomenon. Stars, dust and gas also orbit the Galactic centre but at a faster rate than the density waves, so they catch up with them. As they encounter the density waves the dust and gas is compressed leading to star formation, which is how we can observe these waves. This situation leads to pulses of star formation. These stars continue moving and pass through the density wave and out the other side.

Colliding galaxies can show a massively enhanced rate of star formation and are called starburst galaxies. Despite the collision that produces these galaxies, the stars themselves, somewhat counter-intuitively, do not show much of an increased tendency to collide with each other, as they are still far apart. In fact it is clouds of dust that are colliding, producing concentrations of material ripe for star building.

The gas and dust occupying the spiral arms is called the Inter Stellar Medium (ISM). It is not at all evenly spread but is made up of interacting discrete components.

> ### *The Inter Stellar Medium*
>
> *The ISM consists of clouds of dust and gas, which vary in their density, number count (the number atoms or molecules in a particular volume, generally expressed as number of molecules in a cubic metre), temperature, mass and volume. Throughout the ISM the density is in fact very low, so low in fact that what scientists on Earth would describe as a vacuum is more dense than the ISM, but the sheer size of some of its components means we are talking about very massive objects indeed. The hotter components of the ISM tend to have a low density, which shouldn't be much of a surprise as hot gases expand. Amongst the hotter objects are supernovae remnants. These largely consist of raw atoms and contain few molecules, but are the main means for heavy elements to enter the ISM. They tend to have masses a few times that of the Sun so are not especially large; they also have a relatively low density. As time passes these remnants disperse through the ISM. Other relatively hot components are what are called the hot and warm intercloud media. These are widespread throughout the ISM but have*

> *such low densities that they are transparent to most electromagnetic radiation and are difficult to detect. Diffuse clouds are cooler but can be easily detected at some wavelengths. What is important for star birth are the dense clouds, which are cool and, as the name suggests, relatively dense. Dense clouds have masses of up to ten thousand Suns and some have infra-red objects buried in them, which are very young stars. Diffuse clouds may become dense clouds if there is a mechanism in operation that acts to compress them. Density waves are likely candidates and indeed very young stars are found in the spiral arms, where density waves would have most effect. Also when supernovae explode they generate a pressure wave, which spreads into the ISM and can compress diffuse clouds.*

Whatever mechanism is involved, the density and coolness of the dense clouds makes them the obvious choice to look for stars in the act of formation. The Jeans mass (calculated by an astronomer called Jeans) is the mass that a cloud of a particular size must attain before it starts to collapse under its own gravity. Smaller clouds must reach a higher density than larger clouds to achieve the Jeans mass. If you consider a large dense cloud that has started contracting, inevitably there will be differences of density within it. Now a part of the cloud, being smaller than the bulk of it, will not necessarily have achieved its own Jeans mass at the same time that the overall cloud does. As the main cloud contracts this part of the cloud can attain its own Jeans mass and this is obviously true for other parts of the cloud. When parts of the cloud achieve their own Jeans mass they start to contract independently of the main cloud and the cloud starts to break up giving an overall clumpy sort of effect. As compression continues this clumpiness will also affect other parts of the cloud right down to clumps of the cloud that are the right size for the formation of individual stars, thus a hierarchy of clumpiness results. There will, of course, be parts of the cloud that never achieve their own Jeans mass and, for the time being, will remain as gas and dust.

What happens is this: A diffuse cloud starts to contact after encountering a density wave or a supernova pressure front, which would add heavier elements to the cloud. The cloud becomes a dense cloud and starts to contract under its own gravity. Parts of the cloud achieve their own Jeans mass and start to contract separately from the main cloud so the cloud starts to break up. Within the individual parts of the cloud bits with stellar mass start to collapse by themselves. Stars form out of these bits although some are left behind as gas and dust. The final result is groups of stars embedded in cloud stuff.

The number of stars within these groups will depend on the size of the original cloud and can theoretically vary from one to several thousand. All the stars in a group will have the same age but, depending on the final clumpiness of the cloud, will come in a range of masses, although smaller stars are more likely than large ones. Clusters may be open (widely spread) or closed (tightly packed) and have wildly differing numbers of stars. In star clusters you can see a lot of variation in the number of stars and their proximity to each other. You can also see stars of different mass, often with high mass stars already leaving the main sequence, which enables clusters to be dated.

This echoes the Goldilocks effect for planets. You can have clusters where the stars are crowded together and there are a lot of high mass stars. Such clusters have so much going on in terms of interactions between the stars that any possibly habitable planets would be continually battered. At the other extreme small clusters may not have anything but small stars and little material to build planets from. In between there is the chance for a mediocre star to keep its distance from other stars and have a stable enough environment for planet formation.

The conclusion of the story is that if you want to have a star that supports life (particularly intelligent life), avoid all galaxies that are not spiral. Do not put your star in the central bulge of this galaxy or in its halo. Only put your star in the spiral arms. Within these arms find a moderate sized cloud of dust and gas with which to build your star. Finally make sure that your galaxy does not have a highly active centre. Then compress your cloud and wait for a few thousand million years.

Chapter 10

What About Intelligent Life?

In the previous chapters we have seen that not just anywhere will do as an environment for life, let alone intelligent life. First of all you have to get your galaxy right (i.e. a spiral one) and are limited to the spiral arms. The star cluster should be neither too widely spaced nor too close packed and contain stars of the right age and metallicity. The mass of any possible stars must also be within the right range for stability over a long time span, without being too small to support life. The stellar system itself should contain rocky planets of a similar size to Earth in the habitable zone of the star. If they are much bigger, they will retain so much atmosphere that a runaway Greenhouse Effect is inevitable; if they are much smaller they will lose their atmosphere too quickly for intelligent life to develop. Those systems containing one or more 'hot Jupiters' can be discounted, as in their inward migration they would have swept up any rocky planets. The presence of at least one giant planet in the outer regions of the system may also be essential, as this can act as a shield, protecting the inner planets from too heavy a bombardment from other bodies in the system. Finally, and perhaps controversially, I believe that the presence of a large moon, to both stabilize the planet's axial tilt and support plate tectonics, is a must.

A lot of this chapter looks forward to new observations confirming extraterrestrial life, which could appear fairly soon, probably in the next decade or so. I firmly believe that these observations will find evidence of simple life, as after all, going by our experience on Earth, life can find a foothold in an enormous range of environments, provided enough of the basic materials are at hand.

Most of the methods used to locate extraterrestrial planets are, as mentioned earlier, not yet capable of imaging an Earth size body and direct observation is the only method that could work. New space telescopes, if they ever get off the ground, should give us the capability to obtain such an image. Just a simple detection of an Earth sized planet would give us a lot of information. The metallicity of its star would give us some idea as to what else may be lurking in its stellar system and what sort of materials the planet may be made from. This alone could weed out many unpromising possibilities. Information from the star itself would locate the habitable zone, so we could tell if the planet is there.

The age of a star is reasonably easy to estimate and we can assume that this is more or less the age of the planet. Younger stars and their planets would probably not have had time to develop complicated life and in older ones the planet may have been out of the habitable zone before the star heated up and we observed it. If we can observe an Earth sized planet it means that we can also observe larger planets in the stellar system and so reject systems where giant planets have migrated inwards through the habitable zone. We can also reject systems that have no giant planets at all, as giant planets further out from the habitable zone shield planets nearer the star from incoming comets and other such projectiles. We should also be able to detect the presence of a decent sized moon by the wobble it introduces to the orbit of its planet.

The orbit of a planet can be easily calculated (well, easier than a lot of things) and highly eccentric orbits, which carry the planet outside the habitable zone, can be rejected. We can achieve all this by doing nothing more than taking a few measurements of the star and its planet and back of the envelope calculations.

However, at the distances that we are likely to detect an Earth like planet it will not be possible to see any details. All we will be able to see is the light reflected from the planet, so we will have to resort to spectroscopy for further information.

Spectroscopy

I have alluded to spectroscopy earlier on without much in the way of detail. Spectroscopy is in fact a fine tool for detecting chemical elements. When looking at exoplanets the infra-red part of the spectrum is of especial interest. Planetary bodies are much cooler than stars and reradiate the energy received primarily in the infra-red, while the stars themselves emit radiation at a higher frequency (the Sun has a peak emission in the visible part of the spectrum). These emissions are called black body radiation and are directly related to the thermal excitement of the atoms and molecules making up the body.

The materials in a hot body vibrate more vigorously than those in a cold body and hence emit energy at higher frequencies. Black body radiation therefore tells us directly what the temperature of a body is and is a characteristic of any body not at absolute zero temperature. (Interestingly "any body" includes the human body, which will radiate energy in the infra-red and is why our bodies feel warm). Electromagnetic energy can be emitted or absorbed in other ways too. Within atoms electrons occupy discrete energy levels and can move between them. If an electron moves from a higher energy level to a lower energy level it gets rid of the extra energy by emitting a photon of a specific energy that depends on the difference between these levels. Conversely an electron moving from a lower to higher energy level absorbs photons of precisely the

> *right energy. Another way energy is absorbed or emitted concerns molecules. Within molecules the bonds between the atoms can stretch or bend or waggle at a specific energy and absorb just that frequency of radiation. This situation is very handy for looking at the elements in a planet's atmosphere as important molecules, such as water vapour, carbon dioxide, organic compounds and ozone absorb and emit in the infra-red, where the planet's black body emissions are located. The absorption is the important bit as dark lines in the spectrum, where the absorption occurs, show up strongly in specific positions. Of course these molecules both absorb and emit at these frequencies at the same rate, otherwise the planetary atmosphere would either heat up or cool down. But the radiation from the planet is coming towards the observer and so all the radiation in line of sight is absorbed; when the radiation is reemitted it is sent out in all directions, so in the line of sight more radiation is absorbed than emitted and this shows up as dark lines.*

It is obvious that planetary infra-red spectroscopy can only be carried out in space as otherwise the Earth's atmosphere would interfere. The detection of water in the atmosphere of a planet would be a strong reason for arguing that it was suitable for life. The presence of ozone implies an atmosphere with a substantial amount of oxygen, as ozone is produced by ultraviolet radiation acting on oxygen and it does not stay long in the atmosphere.

All this is quite straightforward but there is a caveat. An Earth sized planet is a very small target if it is even just a few light years away. To give any chance of obtaining a decent spectrum the planet would have to be observed over a period of several weeks, which makes the identification of Earth-like planets a fairly long drawn out process. On the other hand no planets anywhere near the size of the Earth have yet been detected so, unless there is a flurry of discoveries, time to observe one planet for a longish time may be exactly what we have. This is assuming that any such planets exist in near space; my bet is that there will be perhaps a dozen within a hundred light years which are the right size, but I am not betting on there being any intelligent life detected.

Although water vapour, oxygen and carbon dioxide would, in my opinion, categorically demonstrate that a planet contains life, there may well be life on planets that do not have oxygen in their atmospheres. Early life on Earth did produce oxygen, but not in the quantities needed to show up in the spectrum of the Earth's atmosphere. Thus lack of oxygen does not mean lack of life. There are other wavelengths which we could observe. Other chemicals in the atmosphere may be out of balance (from the point of view of an atmosphere without life). For example methane was produced by organisms on the early Earth and indeed this is still happening. Methane has an absorption band in the near infra-red so could be picked up by detectors that will be deployed in the near future, although the absorption is rather weak. It must be remembered however that, as in the case of Titan, there are other mechanisms that can supply methane to a planet's atmosphere.

So what about intelligent life? Of course the only example we have to go on is here on Earth. It is widely assumed that mankind is the only truly intelligent life on the planet, with the possible exception of some whales and dolphins, but they don't have the dexterity to make the ever more complicated tools necessary for communication off planet. Human civilisation has only really been in existence for ten to fifteen thousand years, when people first started to congregate in villages, towns and finally cities. This allowed for increasing specialisation, so some people could get really good at doing things. The concentration of minds, the exchange of ideas and the competition to be the best all spurred innovation. Even with all this it is only in the last few decades, about fifty years or so, that we have developed the necessary means to communicate off planet - and this has mostly been unintentional.

We know that other ages have seen other dominant life forms and it is fun to speculate on intelligent dinosaurs. After all their numbers were dramatically decreasing before the asteroid impact that ended the Cretaceous Period, so perhaps they had developed space technology and were in the process of leaving their polluted planet, which was undergoing rapid climate change. If so they left no evidence behind them as far as we can tell.

The very short time span that humans have been around and the much shorter span during which we have been able to send communications to outside the Solar System, means that it is virtually certain that any extraterrestrial intelligence will have an older civilisation than ours and be technologically more advanced. At the present time the Earth is emitting radio waves that could be detected by an intelligent life only slightly more advanced than ours, but we have only been doing this for about fifty years. This has the obvious point that, for us to detect a reply, any outside intelligence that might have detected and responded to our inadvertent messages could not be further than twenty-five light years away. This is a miniscule distance in terms of the distance between stars.

There is also a limit to how far away an alien civilisation would have to be before our signals are attenuated by distance (the Inverse Square Law again) to become undecipherable. This distance, to be honest, is a matter of conjecture but a hundred light years is a safe bet, albeit a bit optimistic.

We could of course increase the probability of intelligent life finding us by directing a strong thin radio beam with a narrow waveband (this simply means a signal that only occupies a small part of the radio spectrum) towards a suitable target. This was in fact attempted in 1974 via a large radio telescope at Arecibo in Puerto Rico, but only two pulses of two minutes duration were sent and the target was totally unsuitable as it was a globular cluster and globular clusters are made up of old stars with low metallicity.

Although you can blithely say, "Let's aim our telescope at a suitable star", you have to consider what a suitable target is. The best target would be a star with an Earth-like planet in the habitable zone, containing an intelligent race with access to large radio telescopes. Oh, and it should also be close by, so we have some chance of getting a reply. The trouble is that, if we already have the information that defines the best target, we will already have enough information about it, so any message we send would be pointless (except the radio telescope can become part of an interstellar telephone service).

Of course we don't have this information, so there is still a need to select a suitable target as the number of stars that we could aim for far exceeds our capacity to broadcast. Obviously this is also the main theme of this book and the rules for deciding where to broadcast are essentially the rules for the possibility of finding intelligent life. So we look for close (in stellar terms) Sun like stars, without any hot Jupiters, with a gas giant in a similar position to Jupiter, of an age similar to the Sun, located in the spiral arms of the Galaxy, with an Earth like planet in the habitable zone and preferably with a large moon.

Needless to say, if they exist and are relatively close by, such intelligences will almost certainly have detected us first and could be on the way to see us without telling us. At best this is impolite, at worst we should be very, very afraid.

Now we don't have this information about any stars yet, so the best bet is to target stars that satisfy at least some of the necessary criteria. Effectively the approach would be to target nearby Sun-like stars that have not shown any hot Jupiters. This still gives us a lot of stars to choose from, but it will have to do until we have better information.

The next problems we have to face are - what frequency to send the message on, how to code the message so that it can be understood and what to put in the message. I have up to now been assuming radio communication but a laser could also be used. The advantage of a laser is that a very intense, but short lived, pulse could be used. This would be good for saying, "Hi, we are here!" but not so good for encoding messages, at least not with present technology. Radio frequencies have the advantage that we can continuously transmit and the signal is less likely to be swamped by the Sun's output. The main frequency that has been suggested, by Carl Sagan among others, is the frequency of a common emission line of atomic hydrogen (which any intelligent race would know about) multiplied by a universal constant such as π, which should equally be well known by any intelligent being. (It would not work using just the hydrogen emission frequency, as natural emissions would swamp the signal). The precise frequency probably doesn't matter much as, even with our relatively young radio technology, we can scan rapidly across many frequencies, so we can assume that any receiving civilisation would be able to do even better.

An important criterion is to make sure that any signal we send should be obviously artificial. One of the obvious ways to do this is to send a sequence of prime numbers (numbers that can only be divided by one and themselves). There is only one logical way to encode the message, which is to use the binary system (0's and 1's) as this is the simplest number system that can exist. The Arecibo message used 0's and 1's to send a two dimensional image, but the result looked rather abstract and deciding what it was meant to depict proved a problem even to Earth scientists. A larger two dimensional array may serve better to transmit useful information (effectively a television transmission), so some sort of dialogue could be set up, which would require a much longer and more detailed signal, especially bearing in mind that any reply would take decades to receive.

The content of any message is more problematical. We have no idea what any intelligent race might be like and so we could be inviting a swarm of xenophobic maniacs for tea and cake, which could be a disaster if they are like me and prefer cream buns with their tea. This problem is probably intractable. We have already inadvertently transmitted our position and quite likely a lot of information about ourselves. Any further information we send may just be the equivalent of an invitation and your average xenophobic maniac would probably not wait to be invited anyway. Our assumption must be that we are dealing with a more benign intelligence.

In reality I have not found anyone who is suggesting anything more than transmitting basic mathematical and physical data. I say we should wake up our ideas. If these intelligences exist then give them credit for their intelligence and broadcast much more detailed information; if they are worth their salt they will work it out. A single broadcast on a single channel is ludicrous as all you can do is alternate mathematical information, to get a base line for communication, but then we would have to wait for a reply before going on to more complex stuff.

What I am suggesting is having a strong, narrow band channel repeating itself constantly in order to give any intelligence listening information as to how we encode data. This narrow band channel should direct any recipient to a broader transmission (much like broadband use for the internet), which does not need to be such a strong signal. This broadband signal will not need to do the kindergarten bit. We could then just give them a full on television transmission, complete with colour and sound, and let them work it out, as they are after all supposed to be intelligent. Such a broad transmission does not need much repetition as we can send whatever we wish whenever we wish. The content should cover basic facts of Earth life, such as that we are oxygen breathers; we have two sexes; we have a basic moral code that most of us believe in; that technologically we have the potential to move out into the galaxy and we share our planet with a multitude of other life forms related to us. Obviously we should send our peaceful intent, which automatically makes it clear that we would like to set up a two-way communication for our mutual benefit. Bearing in mind the odds against alien intelligences nearby, this may well be the first communication with a different civilisation that they have had as well.

In the field of communication with outside intelligences by far the major effort has gone in to detecting signals directed at us. The Search for Extraterrestrial Intelligence (SETI) is the big player in this game and has been going for forty-five years; it even sparked the fun idea to get people to use the idle time of their home computers to help in the search. The results from SETI are a bit disappointing and the stark fact is that there has been no transmission detected from anywhere that SETI has pointed its radio telescopes at, which really means no communication from all the possible stellar systems that are relatively near to Earth, as these were covered early on.

Logically we cannot say that this negative result implies there are no communicating intelligences in our part of the Galaxy. For one thing there are such a lot of stars to look at. We are only looking at specific radio frequencies, which although obvious to us may not be so to any alien civilisation. Another possibility is that aliens may not want to communicate and their transmissions, even inadvertent ones, are shielded in some way. This does not have to be deliberate shielding but perhaps just a case of economy. Our own radio communications are very wasteful in general because most of the emitted signal does not reach the target receiver. A race that was conscious about waste may well use directed signals more efficiently than we do.

Other intelligence may well use other means of communication. We make limited use of lasers but they are more efficient than radio waves and are easy to send in a narrow beam. Altogether different methods could be used, for example we are just coming to grips with the potential of quantum effects for transmitting information and this method would be almost impossible to eavesdrop.

Although I have suggested that some intelligences may not want to communicate with us I can only see this as an outside possibility. Any race that developed a technology advanced enough to communicate across space would have to be co-operative, at least amongst themselves, otherwise they would have to keep re-inventing the wheel (or their local equivalent). It is possible that aliens are so advanced that they are just ignoring us because we are not worth the effort. This seems unlikely, simply because curiosity must be a trait in an inventive species and, if nothing else, our biological sciences should be different from theirs. From a practical viewpoint, any civilisation that has gone to the effort to avoid detection by us must, for all means and purposes, be considered not to exist because, unless we drop in on their stellar system, they will have no effect on us whatsoever. Altogether SETI should not be written off but it doesn't look hopeful for finding intelligent alien life.

Another question that has to be asked is, "Where are the alien spacecraft?" It seems obvious that a race much older than humans would have already come across the problem of limited resources in their own stellar system, leaving them with two choices, stay where they are and face extinction or build spacecraft capable of reaching other stars. The distance between stars is enormous but this is not an altogether intractable problem for a technological civilisation. The fact remains that, despite numerous UFO sightings and various abductions of obscure people in sparsely populated areas, there has been no reliable evidence of alien spacecraft. Having come all this way I am sure they would at least drop in and say, "Hello".

Our present efforts in space exploration look good so far, with lots of attractive photographs of bodies in the Solar System and beyond, and truly show a rapidly advancing capability, but we really are only just dipping a toe in the water. The problem is not really one of technology but rather politics and a somewhat warped approach to economics. As a race we seem to accept that high technology enterprises have, as a spin off, enormous benefits for the human race, but then quibble about committing what is after all a relatively small part of the Earth's resources to what will ultimately give an unimaginably huge payoff in terms of the continuing survival of our species. What is lacking is a long-term view, which is something that politics seldom even contemplates.

Surprisingly we already have the ability to send spacecraft to the nearest stars, it just needs cooperation and some effort from whoever would fund the project. Nowadays there is a small, but growing, movement towards non-governmental bodies entering into this field, which is likely to improve matters as it removes some politics from the equation. The knowledge to build a viable interstellar spacecraft already exists, although obviously it has not been done yet and requires a lot of resources. First of all a reasonably sized starship capable of carrying a lot of people, about a thousand people would seem right for such a long interstellar journey, could not conceivably be built and launched from Earth. However, there are many iron rich asteroids in nearby space, which could be used as both mines and orbital platforms. Having built the thing we must of course propel it for a very long distance, so we require a very efficient energy source. I would go for a fusion drive as it has so many advantages, not least producing ten million times the energy per unit of mass compared to conventional chemical fuels.

How to Drive a Starship

The most critical part of any starship is its means of propulsion. Chemical rockets will not do at all as they are highly inefficient and you would need to carry far too much fuel, making long journeys impossible. You would also need to find and refine the fuel and it is not at all obvious that suitable chemicals would be available on arrival at a new star system.

You could attempt to go the "Star Trek" way and use an antimatter drive, as antimatter does exist and would be incredibly efficient, as virtually all the mass would be converted to energy. There is a catch however, antimatter has been produced in high-energy physics facilities and on a regular basis, but the total amount is just a few atoms and this has taken vast amounts of energy to produce. As antimatter is immediately annihilated upon contact with matter, it has to be contained by strong magnetic fields and even then does not last long, due to the near impossibility of keeping out every scrap of matter. You may think that somewhere in the Galaxy there is antimatter lying around and we could obtain half an ounce or so for our fuel, but unfortunately no. Antimatter was produced in the Big Bang but was annihilated early on; none is left because there was (and nobody knows quite why) an excess of matter over antimatter from the beginning, so ordinary matter now dominates. We know this because any large collection of antimatter, for example an antimatter galaxy, would be interacting with the matter around it in a predictable way i.e. we would see it if it was there. The only feasible way of powering our starship is by nuclear fusion, the same way the stars are fuelled. I suppose you could consider nuclear fission, where the energy comes from splitting heavy elements, but this produces not only less energy but also a lot more nasty radioactive elements. Not a good plan on board a populated starship. Many scientists are a bit dubious about nuclear fusion, primarily because decades of time and loads of money have been spent trying to build a commercial fusion reactor to produce electricity. So far physicists have just managed to achieve the break point, i.e. getting marginally more energy out than was put in, although the new reactor being built in France

> by a consortium of countries may succeed. These reactors are fusing deuterium (hydrogen with a neutron added) and have shown that it can be done. For a starship engine the whole thing is a lot easier. A reactor confines the ingredients and uses this to heat water to drive a turbine, the difficulty arises in trying to contain the reaction with magnetic fields and everything gets incredibly hot and radioactive. This is the main stumbling block in a ground-based system. For our engine the requirements are different, instead of trying to contain the reaction we would instead use the magnets to direct the energy produced in one direction to gain thrust, a much easier scenario and one we could manage now. Any residual heat could be used to power the onboard facilities. The containment magnets do need to be made of superconductors to be efficient, which means that they have to be cooled close to absolute zero, this is a problem for an Earth based reactor. For our star ship the magnets would be mounted outside and cooled by the cold of space, making the whole machine much lighter and efficient. Deuterium is readily available from seawater and could easily be obtained from any gas giant that we found, so we would have an almost inexhaustible supply.

How fast you could go is a critical question, considering the distances involved. The speed of light has, for the moment at least, to be considered the upper limit that could be achieved.

A feasible aim is to try to achieve half the speed of light by accelerating at one Earth gravity (1G); this would make it comfortable for our astronauts, and is achievable with our fusion drive. At the midpoint of your journey you would simply turn your starship around switch off the engine and cruise at half the speed of light (there is little in inter-stellar space to slow you down), you would of course have to spin the spaceship to maintain gravity. At the appropriate point the engine would be switched on again and then you would be decelerating at 1G, maintaining gravity in the ship and neatly avoiding flying into the target star at a high speed.

Even taking into account unexpected hiccups we should be able to reach the nearest stars in twenty years or even less (achieving half light speed at 1G acceleration would take about five years, so you would have a range of just over five light years). Of course we would not stop at one ship but would send more. Meanwhile our first ships out would utilise materials from the systems that they had arrived in to build more ships. With a Galaxy one hundred thousand light years across we could reach all the stars by this method in a million years. (Interestingly enough, this is about the expected lifetime of the average species, though this probably does not apply to an intelligent species that can avoid extinction using technology.) Such ships would carry enough people to seed viable colonies on suitable worlds, or more likely make worlds suitable for them by terraforming.

So the whole point is that if we, with our very young civilisation, can consider doing this and are within spitting distance of achieving the technology to make it possible, then any alien civilisations slightly in advance of us should be able to do it too. Where then are the aliens and their spacecraft, they certainly are not parked in Earth orbit.

Where Are All The Aliens

Chapter 11

Epilogue

We are obviously at the very beginning in our efforts to find any extraterrestrial life, never mind the intelligent, communicating sort, so it may seem a bit presumptuous to come to a conclusion one way or another about its existence. It is fairly easy to see how single celled life could evolve in many different environments and it seems that almost everywhere we look on Earth – in even the most inhospitable places – we find signs of it. The fact remains that we have yet to find even the simplest kind of extraterrestrial life, although I am confident that we will. Complex life, however, needs a much more stable environment and, arguably, a lot more time to develop.

Astrobiologists are caught in a bit of a cleft stick. Scientifically it is bad practice to take one sample and make lots of assumptions and extrapolations from just this. But astrobiology has only the Earth as an example of a life-bearing planet and so has only the one example to work on. This can lead to the accusation that astrobiology is a bit like religion, in that it places the Earth at the centre of the Universe. However, it's not really as bad as all that because, although we know of no other place where life exists, we do know a lot of places where it definitely doesn't and of course we know that the Earth is by no means the same all over, being full of strange habitats; well strange from our viewpoint anyway.

As our knowledge and understanding of our planet increases it is becoming clear that the complexity of the total environment and how all the different parts interact is vital. The chemistry necessary for life depends on geology, via plate tectonics, which in turn relies on the presence of a nearby large moon. Basic geology, in the size and make-up of the core, gives the magnetic field which helps shield us from lethal radiation. The stability of the sun and our distance from it controls our climate, and its position in the Galaxy means we are less likely to be zapped by a supernova explosion. All these elements, and more, need to come together in a positive way before complex life can survive and evolve.

Although it is impossible to give an actual probability of intelligent life occurring outside the Solar System, and I wouldn't dream of trying, all the evidence so far is rather depressing. There are vast numbers of stars in our Galaxy and until recently the idea of an Earth-like planet around a star similar to the Sun would not have seemed the least bit exotic. Then when our information started to become more detailed everything unravelled. Continuous habitable zones around stars were shown to be narrower than previously thought. Binary planets such as the Earth/Moon system needed such precise conditions as to be almost unbelievable and we know how important the Moon is for continued life on Earth. The dangers facing a habitable planet whizzing around the Galaxy started to become disturbing, dust and supernovae being just some of the most common hazards. What finally clinched it for me was the absurd number of hot Jupiters being discovered. They are everywhere you look (not quite but it sometimes seems like that). As a long-term proponent of the possibility of alien life I have come to despair, or at least be a little unhappy. Now I think the answer to the question "Where are all the aliens?" is simply, "There aren't any".

If there are no aliens out there then it seems to me to be even more urgent that we develop our own starships, otherwise intelligence will prove to have been a failed experiment, once conditions become untenable on Earth. I have outlined how a starship could be powered earlier on and building it with materials from the Asteroid Belt is not an insurmountable problem. One intriguing outcome, if and when we launch such ships, would be the effects on the human race itself. Many people argue that we are no longer affected by Darwinian evolution, as cultural evolution is so much faster, but I beg to differ. Diseases such as AIDS are still affecting our evolution, and we all have genes that mutated about fifteen thousand years ago, at the beginning of agriculture, particularly those associated with the immune system and the ability to digest milk and grains. In fact the evolution of the genus *Homo* (to which we and our immediate ancestors belong) has proceeded more rapidly over the last two million years or so than any comparable animal of the same size. More important however is genetic drift. Mutations are always arising in the population and everyone carries many of them, usually without knowing about it because mutations, despite what creationists argue, are not always harmful but usually have little effect. It tends to be combinations of mutations that are noticeable and even then they are not necessarily harmful.

New species can arise when populations are separated from each other so that they do not interbreed and this is a direct result of genetic drift. The mechanism is easy to understand, as genetic drift is random, so groups of a species that do not interbreed with other groups will be sharing different mutations amongst themselves but not with the other groups. Eventually, enough mutations will occur in the different groups, which prevent them from interbreeding, so new species can arise. In our starships we will be sending off groups of people who will not be breeding outside their own group. Given time these groups will deviate enough from each other to be considered different species. Effectively, rather than finding intelligent races we will be creating them. This is uncannily like Star Trek where, at least in the later series, the various alien races were supposed to have derived from a single species in the distant past.

I bags that I get to be a Klingon.

Where Are All The Aliens

Where Are All The Aliens

Lightning Source UK Ltd.
Milton Keynes UK
01 February 2011
166710UK00002B/81/A